T0301285

Reviews of *m-Profits:*

"Just like the book *Services for UMTS*, this book offers a very interesting perspective on the dynamics of mobile money making. It is particularly interesting because it attempts to deal with the wireless industry at the moment of the sector's global recession and retreat. I am sure the book will be a good read not only for academics but also for industry professionals, operators, bankers and analysts."

Voytek K. Siewierski, Executive Director
Global Business Department, NTT DoCoMo Inc.

"This is the first book I have seen to discuss the money side of the 3G opportunity, with very thorough coverage from the angles of the major players. This book gives practical guidance on creating win-win opportunities to bring content and satisfied users to 3G."

Steve Jones, Founder
the3Gportal.com

"*m-Profits* is one of very few books on 3G services and the only one to give a comprehensive view of the issues of marketing and revenue. It is a useful tool for 3G professionals and amateurs alike."

Sophie Ghnassia, GPRS & UMTS Project Director
France Telecom

"In this book Tomi Ahonen illustrates a solid understanding of how money will be made in 3G with corporate solutions, m-commerce and mobile advertising."

Paul May, Principal Consultant and Founder
Verista; Author of *Mobile Commerce*
and *The Business of E-Commerce*

"*m-Profits* is a detailed, down to earth book and a great guide to the ABC of making money in mobile services."

João Baptista, Vice President
Mercer Management Consulting

"In this book, Tomi Ahonen has made complex theories easy to understand, using practical examples from the leading innovative countries in the world, which can be applied in telecom markets in Europe, Asia and North and South America."

Mark S. Weisleder, Director Channel Development
Bell Distribution Inc., Canada

About the Author

Tomi T Ahonen is an independent strategy consultant on 3G/ UMTS and the mobile internet. A frequent speaker at industry conferences on five continents, Tomi has been quoted over 80 times by periodicals on 3G topics. Previously he set up and headed Nokia's Global 3G Business Consultancy department and set up the segmentation for Nokia Networks.

Earlier he has worked for three operators/carriers in Finland and New York, creating the world's first fixed-mobile service bundle, and setting a world record for taking market share from the incumbent, as well as participating in telecoms standardisation for several years. Tomi has also sold computer networks and services on Manhattan. He started work as a controller on Wall Street. Tomi holds an International Finance MBA (with Hons) from St John's University New York. For more see www.tomiahonen.com

Other books by Tomi T Ahonen:

Services for UMTS: Creating Killer Applications in 3G
Edited by Tomi T Ahonen & Joe Barrett
(John Wiley & Sons, Ltd, Mar 02, 0471 48550 0, 392pp, Hbk, £34.95)

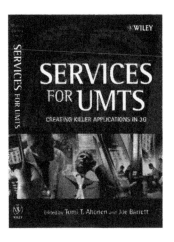

Services for UMTS is devoted to 3G services and provides detailed scenarios for over 170 of them. Written by 14 of the world's leading experts on 3G services, applications, networks and terminals, it discusses the characteristics of mobile services, introduces the 5 M's of how to create value in mobile services and includes chapters on categorising, marketing and partnering. Written for the non-technical reader and with a strong business focus, *Services for UMTS* includes illustrations, statistics, diagrams, analogies from other industries and realistic service vignettes.

Reviews of Services for UMTS:
"*Strong ideas for future demand*" Mike Short, mm02.
"*A must read if you want to understand future services*" Roberto Saracco, Telecom Italia Lab.
"*Manage the $1 trillion bet on the success of 3G*" Assaad Razzouk, Nomura International plc.
"*Explains some of the compelling services in the wireless industry*" Jeff Lawrence, Intel.
"*Insightful discussion into service possibilities*" Dr Stanley Chia, Vodafone USA.
"*Most comprehensive work on the subject*" Regina Nilsson, PwC Consulting.

3G Marketing: New Strategic Partnerships
by Tomi T Ahonen, Timo Kasper & Sara Melkko
(John Wiley & Sons, Ltd, available Nov 02, 0470 85100 7, approx 340pp, Pbk, approx £29.95)

3G Marketing discusses the full marketing and sales side of new wireless services including customer intelligence, segmentation, service creation and management, tariffing, promotion, distribution channels, sales management, portals, brands, reachability, terminals and managing churn.

For more information and ordering details for both books please visit our website at
www.wileyeurope.com/commstech

m-Profits

m-Profits

Making Money from 3G Services

Tomi T Ahonen
Independent Consultant, UK

JOHN WILEY & SONS, LTD

Copyright © 2002 John Wiley & Sons Ltd,
        The Atrium, Southern Gate, Chichester,
        West Sussex PO19 8SQ, England

Telephone (+44) 1243 779777

Email (for orders and customer service enquiries): cs-books@wiley.co.uk

Visit our Home Page on www.wileyeurope.com or www.wiley.com

All Rights Reserved. No part of this publication may be reproduced, stored in a retrieval system or transmitted in any form or by any means, electronic, mechanical, photocopying, recording, scanning or otherwise, except under the terms of the Copyright, Designs and Patents Act 1988 or under the terms of a licence issued by the Copyright Licensing Agency Ltd, 90 Tottenham Court Road, London W1T 4LP, UK, without the permission in writing of the Publisher. Requests to the Publisher should be addressed to the Permissions Department, John Wiley & Sons Ltd, The Atrium, Southern Gate, Chichester, West Sussex PO19 8SQ, England, or emailed to permreq@wiley.co.uk, or faxed to (+44) 1243 770571.

This publication is designed to provide accurate and authoritative information in regard to the subject matter covered. It is sold on the understanding that the Publisher is not engaged in rendering professional services. If professional advice or other expert assistance is required, the services of a competent professional should be sought.

Other Wiley Editorial Offices

John Wiley & Sons Inc., 111 River Street,
Hoboken, NJ 07030, USA

Jossey-Bass, 989 Market Street,
San Francisco, CA 94103-1741, USA

Wiley-VCH Verlag GmbH, Boschstr. 12,
D-69469 Weinheim, Germany

John Wiley & Sons Australia Ltd, 33 Park Road,
Milton, Queensland 4064, Australia

John Wiley & Sons (Asia) Pte Ltd,
2 Clementi Loop #02-01, Jin Xing Distripark, Singapore 129809

John Wiley & Sons Canada Ltd, 22 Worcester Road,
Etobicoke, Ontario, Canada M9W 1L1

British Library Cataloguing in Publication Data

A catalogue record for this book is available from the British Library

ISBN 0-470-84775-1

Typeset in 10/12pt Times by Deerpark Publishing Services Ltd, Shannon, Ireland
Printed and bound by CPI Antony Rowe, Eastbourne

Contents

Foreword

I was deep in thought as I returned from Japan this June. I had been introduced to NTT DoCoMo's third generation Foma services that have been available for close to a year now as well as to the Sha-Mail picture messaging offered by J-PHONE/Vodafone. I was considering the difficult contradiction surrounding the global telecommunications industry. Already functioning in Japan, and being introduced elsewhere, the advanced world of third generation wireless services is a fascinating and imaginative challenge limited in its opportunities only by our abilities to conceive.

However, at this moment consumers seem to have an unwillingness to use, or even a difficulty to use all of the new mobile services and advanced terminals being offered by the industry and the operators.

To describe the market with one word that fits people's lives, that word for the ICT markets today is pain. The foremost reason for the pain seems to be how the industry has become blind to the technology-led way of building the market. The opportunities created by technology in itself have driven to fast, badly timed and, when considered by their magnitudes, perhaps even too massive investments. The day-to-day behaviour and communcation of people and the significance of real needs have been left unseen.

We live in a time of paradox. I do not see any signs around the world of an end to the development of the information society and the digitalisation of various content and services as well as the use of internet-based access, nor that the vision of an information society would be approaching any type of crisis. On the contrary. The development continues strongly but the schedules are different from the guidance provided by forecasts of the companies involved and the industry.

A few weeks after my trip to Tokyo I received surprising e-mail messages from South Korea, which was the other host of the World Cup of football. My friend sat in the new stadium in the city of Suwano and punched up highlights from the game on the keypad of his new colour screen camera phone. And it worked. Inspite of a stadium filled with people, massive amounts of telecoms traffic and other technology, the picture messages arrived very well via e-mail after a little bit of experimentation.

Wireless communication certainly has a bright and successful future. New services and methods of communication based on advanced network technologies and high quality access methods are creating new uses and markets.

As he writes about and evaluates the possibilities and future markets for mobile services, Tomi T Ahonen holds his finger at exactly the appropriate pulse of our times. In fact he knows how to put in proportion the 'pain' felt by the ICT industry and drafts a roadmap away from the pain and ahead for the market. The question no longer is about the role of new technological innovations but rather about how markets are created by innovations around functions and services.

History tells us that no technology alone and by itself has stayed alive and found a place in the lives of people. In addition to possibilities offered by technology, consumers must always identify some additional and real need in themselves, as well as the societal and social context to the services - without forgetting the environment of stories and brands arising from culture and consumption.

I hope that this book by Tomi T Ahonen will stimulate its readers and open the vision to the wireless and mobile communications of tomorrow. It is there after all.

Helsinki, August, 2002
Teppo Turkki
Executive Advisor
Elisa Communications
teppo.turkki@elisa.fi

'In life you throw a ball. You hope it will reach a wall and bounce back so you can throw it again. You hope your friends will provide that wall.'
Pablo Picasso

Acknowledgements

Without Whom This Book Would Not Have Been Possible

There are so many who have had an influence on me and made this book possible. I want to start by thanking Renee Brinker, who taught me to love language; Tapani Leppälä who guided me to creativity; Barry McCauliff who developed my skills in analysis; and Joe Grunenwald who taught me "everything" about marketing. You four set the stage for me to succeed. Thank you also to my partners at the debate teams at Clarion and St John's, especially Dana Murphy and Jeff Lynch; and Robert Abiuso and Mary Colletta respectively. At OCS in New York I want to thank Todd Stevens, Otto Cruz and Aaron Weinstein.

At Helsinki Telephone and Finnet my strongest support was always Matti Tossavainen, thank you. Thank you also Juha Malmberg, Pekka Eloholma, Mikko Lavanti and Gunnulf Martenson. From our team, I want to thank Minna Rotko, Olli Rasia, Tiina Kovero, Nina Lahdesmaki, Tarja Aarnio, Jouko Viitanen, Anne Nikula, Pertti Soitso and Raimo Kirjalainen; as well for guiding us Hannu Peltola, Mikko Heijari, Markku Lempinen and Ismo Heino. We broke records with our product, 999!

Throughout my career there have been a few people who have helped me in understanding the mobile future and its business in more visionary terms. I have been particularly inspired by Teppo Turkki, Risto Linturi, Matti Mäkelin, Matti Makkonen, Harri Johannesdahl, Timo Kasper and Roberto Saracco.

At Nokia I want to thank several people. From early on, Merja Vane-Tempest, Tarja Sutton, Jouko Ahvenainen, Jarmo Harno, Julian Heaton and Stefan Gerrits. Then with segmentation Merja Koistinen, Janne Laiho, Nicole Cham and Russell Anderson. For giving the opportunities and supporting me, Paavo Aro, Aarne Sipilä and especially Ilkka Pukkila. And with 3G, Ebba Dahli, Ukko Lappalainen, Helena Kahanpaa, Arja Suominen, Tuula Putkinen, Michael Addison, Matt Wisk, Spencer Rigler, Malcolm Stout and Matt Taylor.

The work with Nokia's 3G Business Consultancy forms of course the core of my knowhow of the 3G revenues and profits. Thank you Merja Kaarre, Ismo Karali, Kati Holopainen; Timo M Partanen, Vesa Sallinen, Markku Kivinen; Matti Juuti, Krishna Bhandari, Sonja Hilavuo, Maija Gao; Canice McKee, Paolo Puppoli, Rob Hughes, Jari Kovalainen, Kirsten Kuhnert; Reza Chady, Paul Bloomfield; Petra Teranne, Carina Lindblad, Petro Airas, Asko Rantanen, Jaakko Hattula, Paivi Keskinen, Kirsty Russell; as well as Hannu Tarkkanen, Monika Marosfalvi and Harri Leiviska. And we all will sorely miss Tarmo Honkaranta. This book puts in words what you all taught me; in a very literal sense this book would not have been possible without you. Thank you.

I also want to thank all of those who participated in the book *Services for UMTS*, especially Harri Holma and Antti Toskala (Editors of the book *WCDMA for UMTS*) who gave invaluable advice, and of course my writing partner Joe Barrett. In this area I want to thank Mark Hammond, Sarah Hinton and Geoff Farrell from John Wiley & Sons Ltd. for their endless support in both that and this book project.

Lastly but most significantly I want to thank my family, Tina Brans, Jon & Ere Luokkanen; Pirjo Jörgensen; Tepa, Kari, Jukka, Hanna, Jari and Inkeri Lundgren; and Kimi Finell for your support. And most of all thank you to Jan W Brans and Ulla Brans for teaching me to love books, to seek knowledge, to value education. You have supported me at every stage of my life. I want to dedicate this book to you. Thank you.

As someone very dear once told me, I repeat this thought by Gail Mahan: ''When you first see your smile on someone else's face, you've discovered gold. For a friend is the most precious discovery of all. And you'll know that's true when you know a friend has found gold in you.''

Feedback on this book is warmly welcomed, please send it to: mprofits@tomiahonen.com.

Tomi T Ahonen
London 2002

1

'The worst crime against working people is a company which fails to operate at a profit.'

Samuel Gompers

Intro to m-Profits:

Show me the money

The Mobile Internet world is expected to be worth more than 1 trillion (1000 billion) dollars before this decade is over. Most of the services that will generate that revenue do not exist yet. This book looks at those new services with a focus on how money will be made in that new world, and about who will be making the profits.

While the future of mobile telecommunications is impossible to predict in a precise way, broad categories and general service ideas are emerging. Many of the futuristic proposed services are emerging in simplified versions around the world, such as making payments by mobile phone, receiving news, mobile banking, games, etc. As mobile services are evolving fast, a comprehensive description of what is currently available would be out of date by the time this book is printed. So rather, a few illustrative examples of real services are discussed at the end of the five service chapters. For those who would want a deeper discussion of how services are created in next generation networks, please refer to the book *Services for UMTS*, edited by Ahonen and Barrett (John Wiley & Sons, 2002).

This book is about making money with new mobile services for new networks. The number of users on mobile networks has been growing at a phenomenal pace during the last decade and while growth rates are projected to slow, subscription penetration rates are now projected to grow well past 100%, meaning that many people will have two mobile phones. This development is seen in the projection of Western European mobile phone subscribers by JP Morgan.

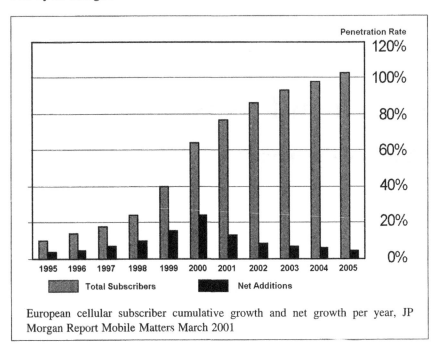

European cellular subscriber cumulative growth and net growth per year, JP Morgan Report Mobile Matters March 2001

The industry on the whole is very young. The cellular telecoms industry started in Japan in 1979, the Internet emerged as a mass-market service through Internet e-mail and web browsers in the early 1990s, and the mobile Internet started in Finland in 1998. The first commercial 3G network went live on 1 October 2001 in Japan. The 3G world is the most powerful mobile Internet environment and in it the telecoms and datacoms worlds combine the speed and dynamics of the growth seen in the fixed Internet and the cellular telecoms businesses in the 1990s. As the industry is growing so fast, more is unknown than known at this stage. Still, this book attempts to look at where the early money is.

This book starts with the mobile phone (cellular phone) as a personal device, and shows how we relate to it. The book then examines attributes of 3G services and examines the concept of micro-payments, and then discusses the Ahonen–Barrett theory called ''The 5 M's of 3G Services'' on how to build value for profitable 3G services. Next the book covers service ideas in five chapters which each describe 3G services in detail, showing also who will be making the money. At the end of the book, there are chapters on traffic patterns, money migration, the marketing of 3G services, revenue sharing and partnering, competition, the 3G network operator (wireless carrier) business case, and the future beyond 3G. In the Appendix, there is a glossary, bibliography, listing of useful websites, as well as a services listing and an index.

1.1 Soup du jour is alphabet soup

While not a technical book, inevitably there will be the assortment of abbreviations in alphabet soup. Let us start with 3G. By 3G, this book means next generation mobile networks to distinguish from first generation (analogue) mobile networks such as NMT (Nordic Mobile Telephone) and AMPS (American Mobile Phone System), and second generation (current digital) networks such as GSM (Global System for Mobile Communications), TDMA (Time Division Multiple Access) and CDMA (Code Division Multiple Access). A close synonym of 3G is UMTS (Universal Mobile Telecommunications System).

Many other terms exist that are close synonyms for 3G and individual readers may be more familiar with some of these terms. IMT-2000 is the initial standard defined by the industry for the high capacity mobile network. WCDMA (Wideband CDMA or Wideband Code Division Multiple Access) is one of the leading standards for 3G being deployed worldwide. CDMA 2000 is the other leading standard and varieties of it include 1XRTT, 1XEV-DO and 1XEV-DV. When this book talks about 3G, for practical purposes any of the above could be substituted, even though technical distinctions exist to differentiate between the terms.

Sometimes the term 'mobile Internet' or 'wireless web' is used to describe the future networks. These terms are much broader than 3G, and usually include 2.5G technologies such as WAP (Wireless Application Protocol), GPRS (General Packet Radio System), HSCSD (High Speed Circuit Switched Data), EDGE (Enhanced Data rates for GSM Evolution) and even may include 2G technologies. For most of the services described in this book, there is no inherent requirement for 3G, and most of the services could be offered on 2.5G networks, and can be called mobile Internet services, or wireless web services, or simply mobile services. One of the key differentiating factors introduced by 3G, is 'Quality of Service' or QoS as was explained by Holma and Toskala in their book *WCDMA for UMTS*. QoS classes are illustrated in Table 1.1.

Table 1.1 UMTS Qos Classes

Traffic class	Conversational class	Streaming class	Interactive class	Background
Fundamental characteristics	Preserve time relation (variation) between information entities of the stream Conversational pattern (stringent and low delay)	Preserve time relation (variation) between information entities of the stream	Request response pattern Preserve data integrity	Destination is not expecting the data within a certain time Preserve data integrity
Example of the application	Voice, videotelephony, video games	Streaming multimedia	Web browsing, network games	Background download of emails

1.2 For whom

This book is intended for the marketing, sales, service creation, pricing, partnership and content managers at operators (carriers), service providers and content providers. It is intended to give a general view of the types of services, revenue streams, profit sharing and other aspects of 3G business. While it may seem like a collection of service types, this book is written too early in the industry to be a comprehensive 'catalogue of services'. It is intended to be a thought-provoking sampling of some of the developments in 3G services and help set up the revenue sharing dialogue amongst the various players in creating successful and profitable mobile services.

1.3 Usage

There are several specialist vocabularies that might come into conflict in a book about a converging world such as 3G. There is a distinct telecoms vocabulary quite different from a datacoms vocabulary, for example telecoms might call something a 'switch' while the datacoms world would identify the same device as a 'call processing server' and so forth. Similarly, there is a clear difference between American telecoms usage and British telecoms usage, for example Americans calling something Toll Free Calls, while the British call the same Freephone calls, and so forth. I have worked in datacoms and telecoms, in America and Europe, and personally use a mixture of the languages. For this book to be consistent, I had to select one usage and try to be consistent in it. As 3G is happening in Europe before America, and the defining and distinguishing aspects of 3G are cellular (telecoms) networks, rather than data networks, I ended up selecting British telecoms usage. I apologise to all American and datacoms readers, but ask that you kindly substitute your appropriate terms when needed. Nevertheless, in practical terms, when I say 'mobile operator', Americans should read 'cellular carrier' and so forth.

Many of the issues in this book relating to new services and their usage may seem silly, strange or against all logic to those of us who are over 40 years of age. Remember that 3G is all about the future, and those who will be living it to the fullest are about age 20 today. If you have teenage or college-age children, observe their behaviour and relationship with their mobile phones and any new services. Never think of yourself as a typical 3G user, but rather consider would your children be interested in any given service – or willing to pay for it.

1 Vignettes from a 3G Future: Music Messages

Not long ago my friends would send me SMS (Short Message Service) text messages to ask me if I had heard the latest song of whatever band was hot at the time. Now it's much better, in that they send a part of the actual song, as a multimedia message. It's a great way to share in the new music, and of course these forwarded sound clips have the omnipresent 'click-to-buy' buttons so that if I like the song, I can have it directly downloaded to the MP player on my phone.

A new promotional tool for the music industry will be the click-to-buy musical ads. They will be in the form of a short 30–45 s section of a current song in technically simple format. The ads would include a click-to-buy button to download the whole song in CD quality on to the mobile phone or other device. The musical ads could be sent to registered fans of the bands, and these fans would then spread the ads through viral marketing. As promotions, the cost would only involve the initial musical messages, as the fans would pay for the cost of any forwarding of the messages. Compared with other ways of music promotion, this method would yield a huge saving in cost, and better than any other promotions, enable immediate call to action.

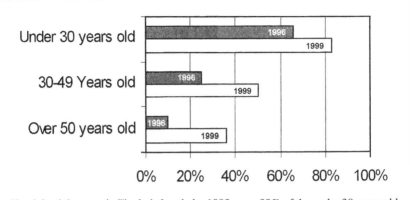

Youth lead the way: in Finalnd already by 1999, over 80% of the under 30 years old population placed most of their calls on mobile phones.
Source Tilastokeskus Three Years in the Information society, July 2001

The youth represent the future especially in 3G and in this book I have tried to bring in latest studies relating to youth adoption and usage of mobile phones. For example Tilastokeskus in Finland has studied how people of different ages place calls on mobile networks and found that of the under 30-year-old group, already 80% place most of their calls on mobile phones. It may seem strange to the older populations, but the youth have always been the first to adopt new technologies and often their behaviour has been judged by their elders as 'a stupid waste of money' – yet they bring their familiarity with new technology with them into the workplace.

1.4 Start me up

This book is a collection of service ideas for 3G, trying to cover them as broadly as possible in one volume, from the viewpoints of not only the 3G network operator, but also the other parties necessary to deliver the service(s). To keep the book in some kind of focus, I have tried to focus on where the money is. Also for those who have read *Services for UMTS*, a similar structure is used, and the two books can be considered companion volumes. *Services for UMTS* answers the 'what' while *m-Profits* answers the 'why' of 3G.

Of course in the final analysis this is a book about the future and is prone to all mistakes that anybody can commit in making predictions about the future. What I hope to accomplish is to bring insight, understanding and inspiration to those who are considering business opportunities in 3G. This is the new world, we need you there. Moreover, it is a vast opportunity where money will be made and where innovation will be the early key to success all over the world. It will not be easy, and there will be numerous failures of individual services. But there will also be phenomenal success stories just waiting to be told. Understanding a new emerging technical standard, and the opportunities it can provide, is not easy. The overall opportunity, however, is so great, that I am reminded of Albert Einstein who said: ''In the middle of difficulty lies opportunity.''

2

'Get your facts first, and then you can distort them as much as you please.'

Mark Twain

Characteristics of Mobile Services:

What makes them different

The 3G service environment is very different from traditional telecoms services, and the 3G environment will be very similar to the fixed Internet mindset in speed, trial and error, and rapid innovation. To put it in perspective, a typical mobile operator in the year 2000 ran about a dozen or two dozen services. These included the basic voice call, the international voice call, the roaming voice call, data access for PC connection, SMS (Short Message Service) text messaging, VPN (Virtual Private Network) numbering plan, friends and family pricing plan, home zone pricing plan, and some more. There really are not very many basic mobile phone services. With WAP (Wireless Application Protocol) the service portfolio easily expanded. But with 3G, the mobile operator will be managing *thousands* of services. As a small indicator, this book has about 180 services in all, and this book is only scratching the surface of the matter.

Building a large selection of services into its portfolio, the 3G network operator has to decide which services to launch and in what order. Some services are not likely to become very successful on 3G. One could build a value chain and partnership model to offer luxury automobiles or houses or sailing boats for sale via the mobile terminal. These are not ideal for 3G, as they are not typically purchased spontaneously in a short amount of time, and the small screen would not make for an excellent presentation media, and mostly these kinds of luxury purchases have many 'touch-and-feel' aspects that are not easily replicated on a mobile phone terminal.

However, there are many aspects of 3G services that make for compelling services. Some of the main characteristics are described here.

2.1 Value in mobile services

For services on mobile networks, the first and most obvious value dimension is mobility itself and aspects relating to movement. Other commonly mentioned value dimensions are personalisation and timeliness, which are not exclusive to mobile networks.

2.1.1 Mobility

By using mobility and the various related attributes such as localisation of content and location awareness of services, the 3G network operator can provide services or content which are better for it. A typical example is the tourist guide such as where is the nearest cash machine, etc. Localisation relates also to the service language when you travel, providing services to you in your own language.

2.1.2 Services should be personalised

The personal attribute for a service makes it feel to the user that it is genuinely unique to that person. If one is a golfer, then services about golf, maps to golf courses, golfing weather, news about tournaments, discount coupons for golfing gear, etc. can be of use. Nevertheless, if that person does not care about tennis, then similar news, advertising and content about tennis is irrelevant. It is very important to build services for the particular interests of the user, and to try to focus the user's interests very precisely.

Many degrees of personalisation can occur. For example if the person likes fashion, then probably fashion news and advertising is welcome. If the person likes Italian shoes and handbags, and the fashion content can be filtered so precisely that not even French shoes or Italian scarves are included, the more personal it becomes and thus more relevant to that user.

2.1.3 Timeliness

Timeliness is another dimension by which services can be improved. We all know from experiences on the Internet how sometimes a 'current' web page may contain out-of-date material, etc. The more timely the information is, the more valuable it often is. Of course, some information does not change much over time, such as geographical entries on maps tend to be relatively stable over time, and having a map update every 15 min would usually not have much added value. Then again, if the map happens to be a traffic congestion map in a traffic congested city at rush hour like Tokyo or Mexico City, then of course the map content could easily have new and very valuable updated information by the minute.

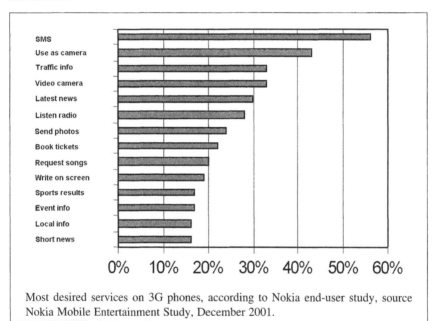

Most desired services on 3G phones, according to Nokia end-user study, source Nokia Mobile Entertainment Study, December 2001.

2.1.4 Time value of information/dilution value of information

With information, there is a strong time dimension to the value of information, which could be called the time value of information. Actually, the more relevant dimension is the dilution of information. The fewer people know something, the more the information has value. In addition, usually the information is spread out over time, which produces the time value effect.

A good example is stock market information. If you happen to catch a private conversation in an elevator that the company next door is about to announce a major patent, and you deduct from that knowledge that the stock price of that company will go up, you might go out and buy the stock. As the information is very narrowly spread, there is considerable value to it. If you told someone, that information would be still very exclusive and it would hold a lot of value. When the company issues a press release about it the following day, then all those who have fast access to such information, such as stockbrokers and those who follow the news feeds, can act upon the information. At that point you still might be able to get a favour from a friend by telling about the press release, but by now the information is with thousands and it has much less value. Moreover, in the evening when the TV news announces it, most of the interested population will have access to the same information. Any significant moves on the stock market will have been made, and as millions know the same information, it holds almost no value.

Services which can capitalise on the fast delivery of exclusive information, and can distribute it while its exposure has been slight, will bring value to those who want that information. 3G news and information services should be built to benefit from this attribute.

2.2 Other attributes of 3G services

There are still many other attributes of 3G services that need further discussion. These are not quite as uniformly applicable to all services but will have many cases where they can be applied. They are multitasking, multi-session, presence and text-to-voice. Most of the attributes for 3G services grow and evolve over time as we go through progressively more advanced systems. A general illustration of the generations was in Kaaranen's book *UMTS Networks*.

2.2.1 Multitasking

Multitasking is doing more than one thing at the same time. Mobile phones allow multitasking for example by walking and talking. You can carry on a phone conversation while you walk from the neighbourhood store to your home. Any places, occasions and situations where time is 'wasted' can be recouped with multitasking. These include all instances of waiting, queuing or temporary delays. In addition, any cases where some monotonous and low-level participation is needed, such as sitting on a bus or babysitting is ample opportunity for multitasking type solutions.

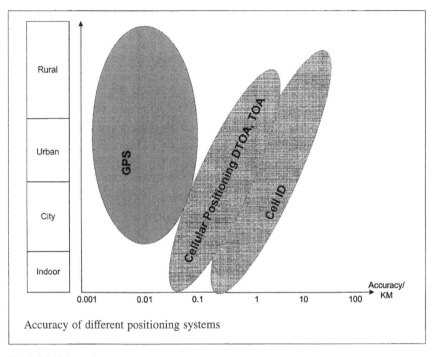

Accuracy of different positioning systems

2.2.2 Multi-session

The multi-session abilities of 3G technology will allow services that are used simultaneously. Most typical will relate to combining the telephone call and viewing something at the same time. A typical example is looking at the movie listings and talking on the phone with a friend who is also looking at the same listings. The combinations are nearly limitless: reading e-mail and carrying on a conversation, downloading a file from the office while paying bills on m-banking, etc.

2.2.3 Presence

Another significant feature of 3G networks is the emergence of presence. Services and utility can be built around presence information. Simply presence means the status of the mobile phone and usually thus its user. In its simplest form, presence already exists in 2G networks, when a person has the phone switched on or off. This can be observed for example when the traveller turns off the phone when boarding an aircraft. We can determine the traveller's status by calling his phone – if the network says that the phone is

2 Vignettes from a 3G Future: Formula One Advertisement and Game

I had the chance to see a live Formula One race at Monaco, and was quite surprised to get the message from Jaguar racing, asking me if I'd like to participate in their racing trivia quiz and be part of their draw for a real Jaguar car. As the participation in the trivia quiz did not cost me anything, of course I participated and in addition to the quiz I also received real-time updates from the race to help me keep track of who is in what position.

Mass events will soon include messages sent on a location basis to participants, often as part of the entertainment of the event. With location information the promoters can target initial ads only to those who are spectators of the event. By offering the contact for free and asking permission to send more ads, the sponsors can gain permission and ensure that ads are not wasted. By combining information, entertainment, mementoes, and games with sponsorship messages, the advertisers can keep the spectators interested in the ads. If carefully designed, such campaigns can yield mobile phone numbers also for future targeted advertising use.

outside of coverage, we know the person is still on the plane, and when the phone rings, we know he has landed. Also sometime networks give different announcements if a phone is turned off, or if it is only beyond the coverage of the network. Moreover, some astute users can detect the location of a travelling person by the ringing sounds when they call the person.

With the advent of 3G presence allows for much more. You can control the device's status remotely by triggers such as time, by authorised persons such as your secretary and by automation such as a computerised device. Presence opens up a vast new area of service opportunities.

Colour screens and imaging phones will integrated camera and video capabilities will help create the early UMTS market. This will start before UMTS with terminals like the Nokia 7650 imaging terminal that also supports GPRS and HSCSD.

2.2.4 Text-to-voice

One of the primary components in multitasking is voice. If e-mails or news are read by voice applications and 'spoken' to the person, then e-mails can be received while doing other things, such as cooking, driving a car or gardening. Many forms of data might be able to be converted to voice translation and delivered in times when reading or viewing is not possible or even desirable.

2.3 Service formulae

There are some convenient acronyms which have been developed to describe mobile Internet applications. The best known are PAIR, MAGIC and 0-1-2-3.

2.3.1 PAIR

PAIR works well to describe consumer services on a mobile device. The letters stand for Personal, Available, Immediate, and Real time. Creating a service, which provides one of these attributes, makes for a potentially successful service. Adding more of the four makes the service ever more useful. Personal was described above, but Available, Immediate and Real time may seem like synonyms, so a few words are worth spending on describing their nuances.

P – Personal
A – Available
I – Immediate
R – Real time

Available means that the information or service can be reached. For example, if you are making sudden changes of plans and all at once need to find exchange rates from one monetary currency to another, then it does not do you any good that the information is on your laptop computer if that computer is not immediately with you and available. But if the service is on your mobile phone, you can access it right then. That is what availability means. Of course, available information may also be immediate and real time, but it does not have to be. Information can be available and distinctly not immediate nor real time, such as the availability of yesterday's newspaper.

Immediate means that the information or service reaches you when it is relevant. For most things, you do not want immediate knowledge. If you do not follow the German Bundesliga (soccer) then you probably do not want goal-by-goal updates of what is happening in the various games played on a given day. So too much immediacy can be a bad thing, we could be bombarded with thousands of trivial bits of information updates. However, if you work in politics, and the opposing party issues a press release, you would want to know when it happens. That is immediacy. There are some things that you are interested in so much that you would want to know when it happens. But you do not want to be bombarded by 'everything' as it happens. Most of the news you probably want to digest at your leisure for example in the morning or at an idle moment during the day, etc.

Real time is the most up-to-date version of the information that exists. Typical of many Internet information searches is that you receive information from a memory cache somewhere, where the information is easily hours, often even days old. This is usually not a big problem if you are searching for Latin phrases or other information which does not change on a daily basis, but on stock market information, the real-time need is quite obvious. You do not want to make your investment decisions on yesterday's stock price information.

One should keep in mind that PAIR is not specific to mobility, so the formula can also be used to generate successful services on the fixed Internet.

2.3.2 MAGIC

NTT DoCoMo have suggested that services have to be MAGIC. Mobile, Anytime, Globally, Integrated, Customised. The MAGIC formula provides a good indication of migrating fixed Internet services to the mobile Internet, or creating new services on the mobile Internet.

M – Mobile
A – Anytime
G – Globally
I – Integrated
C – Customised

One can argue whether all services must be global or globally aware, and to what degree integration is needed for all services, but this acronym provides considerably better mobile service oriented guidance than PAIR. Definitely, mobility, timeliness (Anytime) and personalisation (Customised) need to be considered for a successful service. One could argue that Mobile and Globally are actually part of the same issue of Mobility or Movement, but that is partially a semantic argument of definitions.

MAGIC is another good and simple to remember formula to use when considering how to build services in the mobile Internet space. MAGIC is more specific to the mobile networks than PAIR.

2.3.3 0-1-2-3

Ericsson has offered the 0-1-2-3 system for designing services that are intended to work on a mobile phone. The numbers mean, 0 written manuals – all services must be simple enough to be learned on the phone with online help; 1 button to the Internet; 2 s maximum delay waiting to access any service; and 3 keystrokes is the maximum to gain access to any service or feature.

0-1-2-3 is a useful system with a focus more on the handheld terminal and its interface, but provides also good guidance for service developers. In the book *Services for UMTS*, Joe Barrett and I offered a comprehensive mobile service creation theory, called 5 M's. The 5 M's will be discussed in its own chapter later in this book (Chapter 5).

2.4 At last on attributes

With abilities to take advantage of location information, do multi-session work and provide timely services, new and exciting content will emerge. The attributes described here are useful, but a holistic approach specific to 3G is developed later in this book, where the theory of the 5 M's of 3G service attributes are described (Chapter 5). Before that, this book will examine in more detail two other aspects of mobile services and 3G. First a chapter looks at how important the mobile phone has become to us (Chapter 3), and then the chapter on micro-payments will describe what is probably the single most significant factor in why the mobile services will be a success (Chapter 4).

Still these attributes in this chapter can serve as launching points into new services, to begin to understand the environment of services of the future. The mobile services opportunity is immense. Those who understand what makes mobile services different from all existing services will be in the best position to capitalise on the opportunity. As the US Supreme Court Justice, Louis D. Brandeis said ''Most erroneous conclusions are due to lack of information than to errors of judgement.''

3

'This "telephone" has too many shortcomings to be seriously considered as a means of communication. The device is inherently of no value to us.'
Western Union internal memo, 1876

Mobile Phone the Most Personal Device:

Cannot live without it

This book is not about what mobile phone handsets will be like in 3G. However, to access mobile services, our device will most often be a mobile phone. Most users will not differentiate between their phone or handset and the service they consume. Therefore, it is worthwhile to discuss our feelings towards this relatively new gadget in this chapter.

All through time, man has tried to invent convenient devices to make life better. In addition, as the utility of a given device has become obvious, man has tried to miniaturise the device or gadget to be able to carry it with him. No device that has been adopted by the masses has achieved such a personal attachment as the mobile phone today. But to understand how intense this new attachment to the mobile phone has become, first let's examine the previous most personal universal gadget, the wristwatch.

3.1 First universal gadget – the wristwatch

The obvious most popular personal gadget of all time has been the wristwatch. From sundials to mechanical clocks to pocket watches to the wristwatch, devices to tell time became status symbols, worn by the wealthy and made of precious metals. Eventually after mass production made it possible, wrist-watches became universal devices and most today are made out of plastic and simpler metals. It is the only device that man today straps on to himself and keeps on him often 24 h a day. Nevertheless, it took centuries for the mechanical watch to evolve from a luxury of the few to something that every-body wore. Those who remove the watch at night, tend to dress it upon them-selves first thing in the morning. In addition, of course, many have watches that are waterproof and wear them in the shower, etc.

A few other devices have also reached very close affection with humans. The pen or writing instrument is one which many wear on their person, in a pocket or a woman's purse. Like watches, also fine pens are made of gold and silver, with other rare materials, etc. Again, most people have dozens of free give-away pens and other relatively cheap plastic pens and would not lose any sleep about losing any single pen. But the pen is not as vitally needed as the watch. The pen is not carried quite everywhere, for example not taken into the bedroom, bathroom, etc.

3.1.1 Other contenders and pretenders

Young people seem to be plugged into music players/headphones and porta-ble gaming devices. But of course, most do not actually carry them all day; rather they use them many times per day. When not needed, the music player is usually not carried everywhere. Moreover, the need to keep the music player on all the time diminishes with age. Therefore, the Walkman, minidisc and MP3 players are not as prevalent as the watch.

Portable games are another gadget which some carry all the time. These tend to be restricted to boys and young men. The portable gaming devices are also not carried everywhere, except maybe when a new game is being played for the first few days.

The PDA (Personal Digital Assistant) is another gadget many are passio-nate about. Its users swear by it and are quick to debate the merits of their preferred PDA device. However, PDA penetration is still in small numbers even among business users, even less so when examining the general public. The worldwide penetration of PDAs was at about 30 million worldwide at the end of 2001, while mobile phone penetration was nearing 1 billion at the same time. When even in America the PDA penetration is well less than one for every 10 people, this hardly is a universal gadget, yet.

3.2 Science fiction is here today

Science fiction has promised us fancy gadgets, from the telephone in the shoe used by 1960s comedy spy Maxwell Smart, to the wristwatch videophone used by detective Dick Tracy. James Bond proved the utility of private personal text messages to us in the 1970s movies when he received text messages on to his watch. The original Star Trek TV show of the 1960s promised us handheld location and communication devices ('communicators'), and the very sophisticated handheld computers, recording and measuring devices ('tri-corders'). As Star Trek went into the Next Generation in the 1980s, so too did the communicators, which then had them as voice-activated button-sized location-aware communication buttons. And Star Trek's computer controls became touch-sensitive flat panes.

3.2.1 They all exist in portable devices today

Currently all of those devices exist in some form in handheld type, often of greater ability than imagined by science fiction a few decades ago. Handheld

Mobile Video will be one key UMTS service and a number of products are already being developed with this in mind including this fully integrated video phone from Origami

GPS (Global Positioning Satellite) positioning systems let us find our location exactly. Pager text messaging, and SMS (Short Message Service) text messaging has given us the ability to receive (and send) text-based messages. PDAs are of course very sophisticated computers, and with their plug-in units easily could replicate most of the features of Star Trek's tri-corder and do much more.

Maxwell Smart's shoe phone can be created with current technology, although current digital cellular phones are much smaller and more convenient and fit into the pocket rather than shoe. The various touch screens from assistance computers at shopping malls, to the pen-computing screens in PDAs, have brought us touch screens. Small TV screens have been introduced into various devices, including the wristwatch, bringing part of Dick Tracy's wristwatch video phone; now several 3G concept phones have illustrated similar devices with video telephony. In addition, while the Star Trek button communicator is still a few years away by current technology and miniaturisation, a variation of that is seen in the increasingly popular hands-free headsets of ear microphone and tie clip type microphone, which give us nearly the same communication utility as the communicator button of Star Trek. Ericsson was among the first to make the earphone and microphone wireless by using Bluetooth, so that the mobile phone could be in the pocket or purse or briefcase.

Now with many devices and abilities converging, very soon many of these devices and abilities will converge. One such device is already the Nokia Communicator and its clones, combining in a somewhat cumbersome sized mobile phone also a computer, Internet access, and numerous PDA features and plug-in units. Ericsson has a similar device and many PDA manufacturers are approaching the same objective from the side of adding mobile phone modules or functionality to the PDAs. The personal universal communication and computing device, as seen in visions of science fiction, is here today.

3.3 Mobile phone is the most personal device of all time

The mobile phone emerged in the late 1980s and by the mid-1990s it had gone from being the latest executive toy and expensive gadget, to a mass-market device where many countries had mobile phone penetrations of 30–40%. At such penetrations, most households started to have mobile phones. As the penetrations reached 60–70%, it meant that many children aged 10 had mobile phones and most working aged adults one. By the time the

3 Vignettes from a 3G Future: Coupon from a Friend

These forwarded coupons are pretty cool. There are a couple of friends who seem overly eager to forward them, probably so that they get their maximum forwarding points, but from most friends I know if I get a forwarded coupon, it is for something I really will appreciate. I notice that I am saving lots of them and have been very amused to use them at the stores for sometimes considerable discounts. Luckily since I carry my phone always with me, I am never stuck in the situation that the coupon I want is at home.

The 'permission push' advertisements work best when the actual advertisement received addresses a real and personal interest. Combined with location information, the targeted permission push ad becomes an extremely powerful and activating marketing tool, and one which is almost totally unavailable in any media currently. The real power, however, comes from viral marketing. The registered or known users and customers form only a small part of the potential users, and in very many cases one user knows others who may be interested. By forwarding ads, these people are also reached.

penetrations reached 80%, several of the working aged adults started to have second, even third mobile phones.

The mobile phone has rapidly become the most personal device we have. Not only is it one we carry with us all the time, and one which we personalise with interchangeable covers and personal ring tones. The mobile phone is often used to communicate our personality, it is also one which is starting to replace another extremely personal device – the wristwatch. Some anecdotal evidence suggests that there are some people already who are abandoning the wristwatch in favour of the mobile phone – since most current phones also include time display.

The mobile phone has developed remarkably fast to include many features and abilities, from storing numbers and names, to keeping track of calls, to adding schedulers, calculators, to-do lists, etc. As we incorporate more and more of our personal life into the mobile phone, it becomes also a strong key into our lives, and the heavy users tend to guard heavily the information on their phones. With ever faster speeds the mobile network can enable ever more complex services. The evolution of the speeds involved with mobile services was discussed by Holma and Toskala in *WCDMA for UMTS* and the table here illustrates the speeds involved (Table 3.1).

As the services and abilities are added to the mobile phone, its value to us becomes ever greater and we start to guard it for its personal nature. For example, it is very easy to see who has called you recently or who you have called, or whose calls have you not taken. Even more potentially revealing is your calendar/scheduler where there can be very many personal notes and commentary. The nicknames that you might assign to some people in your phonebook are another case of potential embarrassment. In addition, most revealing of all often is the history of SMS text messages that you have stored, in both sending and receiving SMS text messages. For girls and young women the mobile phone has very clearly replaced the traditional written diary and much of that tradition of writing what has gone on in one's life is now kept in the storage of the mobile phone.

3.3.1 Will not give it up

Another sign of how personal the mobile phone is is that we are not willing to give it up. For most working people the two most valuable daily tools tend to be the PC (especially usually e-mail) and the mobile phone. Depending a bit on how much the worker is desk-bound and thus could have easy replacement via the fixed phone line, but in many cases it is a toss-up which is needed more, the PC or the mobile phone, for conducting daily business. Certainly if

Table 3.1 Network performance

	Macro cell layer	Micro cell layer
Capacity per site per carrier with one power amplifier	630 kbps	—
Maximum capacity per site per carrier	3 Mbps with 3 sectors	2 Mbps
Capacity per site with 3 UMTS frequencies	9 Mbps	6 Mbps
Initial sparse site density	0.5 sites / km^2	—
Maximum dense site density	5 sites / km^2	30 sites / km^2
Maximum capacity	45 Mbps / km^2	180 Mbps / km^2

we combine all phones and contrast that to the PC, the phones win out as (mostly) the most important daily tool at work. Yes, information processing is important, but communication – and the ability to be reached – is more important. You can work on your next presentation temporarily also with paper and pen, but you cannot yell loud enough to reach your colleague who is travelling in France. Moreover, as the newest mobile phones introduce the ability to send and receive e-mail, then definitely the mobile phone wins out over the PC as the single most important instrument at work. A widely quoted early study by Goldman Sachs in the US, where travelling laptop PC users were given the chance to try the Blackberry device, showed a clear preference of using the handheld device over carrying the laptop PC. Twenty per cent of the users were reported to have stopped using their laptops altogether. Certainly it is too early for definitive conclusions, but the early evidence suggests that when given the choice, users will prefer portable pocketable devices over the laptop PC, as long as their primary communication needs are met.

Outside of work there are many gadgets which have exceptional meaning to us. For some it is the TV and its related DVD and VCR, etc. gadgets; for others the stereo-Hi-Fi, and for someone else it might be kitchen devices, if cooking happens to be a particular passion. Again, if we were going on a sudden trip, most probably we would not pack a portable TV. We might well take the portable CD player, but again probably not the microwave oven or the kitchen blender, etc. But the mobile phone would go automatically with us almost everywhere. Again, for those who have grown accustomed to it, they can hardly imagine living without it, or if they do, it is a deliberate decision to be out of contact for example during a vacation.

3.3.2 So what

The point may be well accepted that the mobile phone is the most personal device we have, it has reached that position of being always with us like the wristwatch, and we consider it the collector of valuable personal information – even often view the mobile phone as the very extension of our personality. But so what. For the company considering launching mobile services, or for the company currently having services for other platforms and technologies and networks, this has profound impact.

First, the mobile phone is the only device, apart from the wristwatch, that 'all people' carry with them at all times. For any service, in any country, if your target audience needs any device to consume your service, the person will also own a mobile phone. Even if you serve your customers through the TV, or the Internet, any of the people who own a TV or PC – and have disposable income to consume more – will also own a mobile phone. But the reverse is not true. Many more mobile phone users exist already than PC users, and during 2002 the worldwide mobile phone penetration will grow past worldwide TV penetration!

The mobile phone is the only device apart from the wristwatch that we carry with us every day. Nevertheless, quite apart from the wristwatch, the mobile phone can access many services on networks, and communicate with other devices. The sheer numbers of mobile phone users – and mobile phones – is astounding. The global population of mobile phones is expected to reach 1 billion early in 2002. Contrast that with about 400 million Internet-enabled PCs worldwide. PDAs are still in embryonic stage if considering these numbers, with a worldwide population of about 30 million. For comparison, consider that the worldwide TV penetration is about 1.3 billion.

Secondly, the mobile phone is the first personal device, which has the ability to supplant other things we carry with us. Already first signs exist that some are abandoning the wristwatch and use the clock on the phone.

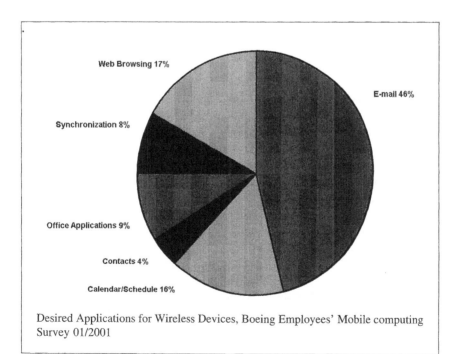

Desired Applications for Wireless Devices, Boeing Employees' Mobile computing Survey 01/2001

Many have stopped carrying a personal address book and phone list as the mobile phone's directory was able to store all significant numbers and names (and in some cases also addresses). With the advanced phones like the Nokia Communicator and other devices like it, some are abandoning the pocket calendar and use the mobile phone as the calendar/scheduler. The benefits are even more pronounced in the work environment. A recent survey by Boeing of its mobile workers illustrated the various needs that the business-to-employee environment has.

The trend will continue as mobile phones add more features. With the advent of mobile commerce and the m-wallet, the mobile phone will start to supplant first cash and then credit cards from our wallets. The built-in snapshot camera will replace the snapshot camera needs and the built-in music player will replace the Walkman/CD player/MP player. Few people want to carry a large load of gadgets, paper booklets, cash, credit cards, leather wallets, pens, etc., if one device can do all or most of that.

3.4 Near future evolution of mobile phones

We can rely on several trends to give guidance on what lies ahead for our favourite gadget. There will be miniaturisation, more convenience, more features, and integrated components and functionality.

3.4.1 More features and functions

With miniaturisation we will be seeing phones which incorporate features from our favourite other portable devices. Many early signs of that already exist, with many mobile phones including some games. Some include radios and/or MP3 players. Others include simple PC-like functionalities and PDA-type functions. Of course, many provide some type of access to the Internet. The first mobile phones with built-in cameras have been seen in Japan. Therefore, whatever is your personal favourite portable device or gadget, even mobile phones of today provide probably some of that functionality, and with miniaturisation, soon most phones will have such features.

Most of all the mobile phone of the near future is a multi-purpose personal device. It will evolve and grow, so that soon expensive top-of-the-line options will become standard features. Much like over the 1990s the PC went from single-use DOS-based systems that were usually equipped to do only word processing or only spreadsheets, to Windows-based multi-use platforms. As we know, PCs today come with standard ability to do not only word processing and spreadsheets, but also with the ability to make presentations, send and receive e-mail and access the World Wide Web. And all along the PC has evolved in ever more compact sizes and with ever faster processors and larger data storage abilities and more built-in components such as modems, sound cards, etc. The user interface will get ever better, evolving just like the early PC evolved from cumbersome DOS-based operating systems to the Windows (and Macintosh) operating systems and now the Internet 'click to go' philosophy. Mobile phones have just taken the jump to colour displays in 2001, and many advanced operating systems and interfaces are soon upon us.

3.4.2 Cost differential will shrink

A similar evolution awaits the mobile phone. Five years from now mid-range and even low-end 3G phones will be equipped with cameras, MP3 players, a lot of data storage size, etc., all in a phone similar in size – or still smaller than – today's GSM (Global System for Mobile Communications), TDMA (Time Division Multiple Access) and CDMA (Code Division Multiple Access) phones. And while 3G phones initially will cost more, especially those with expensive additional gadgets built in, the prices of those will also come down.

Within a few years the cost differential will become totally marginal, and the majority of phones which are sold will be 3G phones. In addition, those phones will be very capable communication and data handling devices. Those phones will replace the current mobile phone that almost every person in the Western world, and an ever-increasing part of the rest of the world, carries today. This growth in the technical ability of the mobile phone will have profound effects on the array of services that can be provided. An even more powerful aspect will be that the 3G phone will eventually replace the PC as the dominant way to access the Internet. That development is discussed in Chapter 4 in this book.

3.5 Nobody does it better

The mobile phone has become a personal device we cannot live without. It is the only communication and data device we carry with us all the time. It does mean that we assign ever more value to our mobile phone, and to the services and features we have on it. The 3G mobile operator has to keep in mind that in this competition the operator has to remain the preferred partner for the phone user. It means innovating.

This book will discuss various ways to do so in the five chapters on service ideas. Many of those ideas may seem strange, even silly, to people who are the target audience for this book. If you feel that way, remember not to think about how *you* would feel about this future, look at how your *children* – or if you are over 50, how your *grandchildren* are reacting to the new. It is irrelevant how silly you and I might think some of the new services are; it is relevant if the young people of today want them and use them. They are the Nintendo generation used to small pocketable gadgets and tiny screens and miniscule keyboards. And the youth of today will grow to be the adults who consume mobile services all their lives on 3G networks and beyond. The old-timers will retire to reminisce about the good old times when we were not continuously interrupted by the chirping of the mobile phone. Focus on the youngest adopters and remember what John Nuveen said about age: ''You can judge your age by the amount of pain you feel when you come in contact with a new idea.''

4

'Anything worth selling, is worth selling twice.'
Ferengi Rules of Acquisition, from the
TV series Star Trek Deep Space Nine

Micro-payments:

The magical key to content revenues

About 100 years ago Alexander Graham Bell said he could see a future when every home would have a telephone. While he was acknowledged as a genius, this thought of his was broadly dismissed as utopian. Early on in his career, Bill Gates said he wanted a PC on every desk. This too was seen as a lot to ask for in a world where it seemed that most offices were happy with the electric typewriter, even as many saw early benefits out of spreadsheet and word processing applications to specialised fields. A few years ago Jorma Ollila said he wanted the Internet to every pocket. That, quite a far-reaching proposition, does not seem that implausible considering how broadly the Internet and mobile phones have spread.

Now I offer this vision of the future: ''Within a few years, all content creators will be paid for their content.''

That is quite a bold statement when we think of the billions of Internet pages on millions of websites, most of which are available for free. Today most content providers are struggling with the dilemma of tiny value. Most websites have a loyal collection of users who clearly appreciate the content – loyal visitors and browsing customers who keep coming back – yet nobody seems to be willing to pay for access to the content. The actual value of a typical web page could be estimated and it is likely to be pennies, or even fractions of a penny, but the current fixed Internet has no given common mechanism to track, bill, and collect sums of such tiny value. While some exceptions exist of pages that users are willing to pay dollars or even dozens of dollars to access, whereas most websites rely on their revenues coming from web advertising, a struggling business which exposed its flaws during 2000 and 2001.

This chapter goes to the heart of the mobile services opportunity, the inherent ability to charge for all content, especially for those requiring payments of less than a dollar, or micro-payments. This chapter will examine the inevitable shift in the near future of access devices in what I call the 'Fundamental Curves'. This chapter will contrast the billing and charging abilities of the fixed Internet to those of 3G network operators. The chapter will explore user attitudes towards being billed, and come to the inevitable conclusion that the best content will migrate to the mobile Internet. The good news is that users on the mobile Internet are also willing to pay for their services. That is why I see that the future holds relief to all content providers: they will be paid.

I want to point out that when content is billable on the *mobile* Internet, it does not extinguish the free content on the *free and fixed* Internet. But just like we can purchase a daily newspaper today and read the Garfield cartoon in it, or we can go to the public library to read the paper for free, so too can billable content and free content coexist in the mobile and fixed Internets. The two are not mutually exclusive, not now and not in the future. In fact early markets such as Japan and Finland prove this point very well. But we have to start with the inevitable trends that are bringing a dramatic change in the population of access devices to the Internet. In addition, to understand how that can happen, let's examine how the content evolved on the fixed Internet.

4.1 How the Internet evolved to adjust to access devices

Many late at becoming web users might not know that only 15 years ago the primary access to the Internet was from university and government *main-frame* computers. The PC existed back then but in the late 1980s, the personal computer was still learning to talk to local area computers in what are called LANs (Local Area Networks) typically from 5 to 50 computers in size, such as one floor of an office. While the Internet had been around and PC-oriented terminal emulators and Internet protocol software such as FTP (File Transfer Protocol) and IRC existed to access the Internet, very few PC users actually used the Internet. Even the early modem sales to

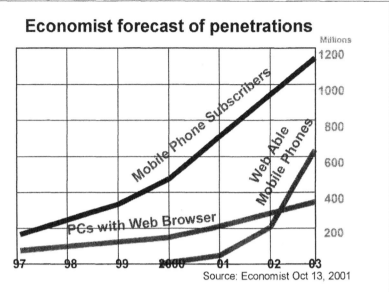

Economist forecast of penetrations

It is now widely recognised that the number of web enabled mobile phones will overtake the number of web browser PCs. In some countries like Japan and China, the first experience of browsing the Internet will be from a mobile phone

PCs during the late 1980s and early 1990s was to provide access to so-called BBS service providers to gain access to bulletin board services. This was all before the first web browser. Almost nobody talked about the Internet, and IP (Internet Protocol) was just a curious, almost obscure government and education-related computer networking standard.

4.1.1 A system optimised for research

The point to remember is that when most access devices were mainframe computers used by universities and the government, for research, then the content and interfaces and use all evolved to optimise the system for their access devices. Any user of Gopher will fondly recall how much faster and more efficient and so much more *organised* it was than the way the Internet is today with web browsers. It was easy to find information. Of course there was no advertising. There was no commerce. The Internet in 1990 was a government and university research tool designed to organise information. Its content and applications reflected that. When new or improved applications and services were introduced, they were built to the main-frame computer environment, and personal computers used 'emulator software' to emulate – i.e. mimic – mainframe behaviour to conform to the existing prevailing environment. A PC in the 1980s had to make itself seem like a mainframe computer (or its terminal) to be able to connect to the Internet. The emulation software and its protocols were tedious to learn and use, and few PC users actually bothered to try to access the Internet at the time.

Why the World Wide Web (WWW) was said to be inferior to Gopher:
Gopher was organised,
WWW was chaotic
Everybody knew how to use Gopher
Gopher was much faster
Gopher content was all ready, but hypertext content had to be rewritten
Nobody needed graphics

With ever cheaper modems, faster personal computers, local area networking and the graphical user interfaces of the Macintosh and Windows operating systems, personal computers became *viable* as a mass market access device to the Internet. But even that was not enough. It was not until there was an easy-to-use Internet access interface – what we now call the web browser – that Internet access became reality. The first web browsers Mosaic and Netscape were later followed by Microsoft's Internet Explorer and finally the PC and its users were able to conveniently access the Internet.

In a very short time, the sheer volume of personal computers dwarfed the population of the first Internet devices, the mainframe computers. And content providers started to design their data and web pages not for main-frame access, but rather for PC access.

4.1.2 Internet evolved to accommodate predominant devices

The Internet content and nature evolved to reflect the newcomers. We still have all of the original content, as the Internet is an excellent research tool even today. However, the Internet has also become an entertainment vehicle. It has become an electronic distribution channel. It has advertising. And it has a lot of commerce. But most of all, if any developer is thinking of how his web page will look, the developer will first focus on the IBM compatible PC population, running on Windows using Explorer or Netscape as its browser. And probably the developer will not spend 1 min of time worrying how the web page might look on a mainframe computer – the original tenants of the Internet.

The lesson we learn is that if a larger population of access devices emerges, the content providers will naturally migrate their content to suit the predominant access device. Moreover, this underlying premise is about to shake the dominance of the PC as the access device, within the very next years.

4.2 Fundamental curves

The worldwide growth in PCs is still growing, but the growth is slowing down. Even if year 2001 sales do not significantly exceed previous sales, the overall worldwide PC population is expected to grow at healthy rates over the next few years. New PCs, both desktop and laptop are all Internet-enabled, and as older generations of non-Internet-enabled PCs are replaced with modern ones, the whole PC population is becoming Internet-enabled. It is estimated that the worldwide Internet-enabled PC population will be about 400 million during 2002.

Global mobile phone penetration shot past PC penetration many years ago. The 1G and 2G mobile phones could not access the Internet. The first crude, slow and simple Internet-enabled WAP (Wireless Application Protocol) phones entered the marketplace in mass only in 1999. With Japanese I-Mode phones, GPRS (General Packet Radio System) and HSCSD (High Speed Circuit Switched Data) the worldwide Internet-enabled mobile phone population is fast approaching that of Internet-enabled PCs. With WAP 2.0, HSCSD and GPRS, the speeds and convenience will improve for mobile access to the Internet just as early web access methods for PCs improved rapidly in the early 1990s. All major analyst's forecasts of mobile phone penetrations project the crossover point in time, in the very near future, when there will be more mobile Internet-enabled phones worldwide than there are Internet-enabled PCs. Many forecasts have this happening during 2002, and some of the optimistic predictions have that happening before this book hits the bookstores.

4.2.1 More people will access the Internet from mobile phone than PC

Therefore, within a few years more people will be accessing the Internet from a mobile phone than from a PC. The sheer numbers of the terminals will dictate this. There will, of course, be very many people with both means, and they are likely to use both – meaning that sometimes you might access the

Internet from your PC, at other times from a mobile phone. However, as mobile phones cost only a fraction of the cost of a PC, the real mass market, when considering a billion web surfers, will come via mobile handsets and other much cheaper devices than personal computers. Yes, Bill Gates may have achieved a PC to every desk – but that is a far cry from an Internet device to every person.

Some sceptics might still argue that mobile phones will reach a saturation point any day now and their growth numbers will stop. Those sceptics have been proven wrong statistic after statistic. The latest numbers in early 2002 have countries like Finland, Iceland and Israel leading with over 80% mobile phone penetrations and Taiwan already passing 90%; all of this growth is still showing no signs of slowing down. In fact, most Western European countries now project mobile phone penetrations at well over 100%. Let me repeat that. The cellular phone penetration will exceed 100%. If we distribute it evenly it means every baby who can't speak, and every great-great-grandmother who no longer can hear, will also have a mobile phone, and still with phones left over to give to the dog and cat of the family.

In reality the over 100% number means that many working people will have two mobile phones, for example one for work and another for personal use. Current Western European penetration ceilings set the penetration to level off at about 120–130%. They also estimate that only about 85% of the humans will have a mobile phone, but approximately every second person will have two phones. Contrast that with PC penetrations. Even in the wealthiest countries with the highest PC penetrations, US and Canada, nobody suggests that PC penetration will reach 100% of the population. In addition, the rest of the world is not as affluent, so the price differential between a mobile phone and a PC is relatively much greater than in North America. So where e-mail and Internet connectivity has been a key reason for purchasing a PC in the 1990s, in the early 2000s in the developing world a mobile Internet-enabled cellular phone will fill that need at much less cost.

4.2.2 Mobile Internet is where the mass market will be

When the crossover point occurs, and more people will have access to the Internet via a mobile phone device than via a PC, then the mobile Internet will be the one with the larger total reachable audience, and content providers will have one compelling reason to design content for the mass market: mobile Internet.

This is actually already happening in some segments around the world. In Japan for example, the PC penetration is relatively low. With university students, the proportion of those who have a PC is low, but practically all students have a modern mobile phone, which today regardless of network will inevitably have Internet access. So professors are turning to using the mobile handsets as a vehicle to share information, hand in homework, make scheduling announcements, etc. Therefore, the crossover point is happening very early in countries where the PC penetration is low, and mobile Internet penetration is high. Soon this phenomenon will also repeat around the rest of the world. Japan is also expected to be the first country where mobile Internet advertising revenues will exceed those of the fixed Internet advertising. This is projected to happen in the early months of 2002.

Nevertheless, the number of users is not the only reason why content providers will prefer the mobile Internet to the fixed Internet.

4.3 Billing on the fixed Internet

When you examine the current fixed Internet, and most content on it, you see several problems to billing for content. First, there is the built-in assumption that all content has always been free on the Internet; therefore, any new content should also be free. This sentiment is a direct cause of the original research-oriented nature of the Internet. It used to be an information-sharing network among academics in government and universities. Students used it to gather information for free. There was every reason to keep the information free. This sentiment has lived long into the PC evolution of the Internet, and content providers are having a very hard time when trying to migrate customers from free pages to subscription pages. Still today most users will strongly avoid any sites that require signing up for the service and pay for its use.

Secondly, the fixed Internet has countless options for similar content. The Internet is such a vast repository of data, and theoretically a complete interconnection between all points of contact on it, that mostly for any site willing to limit access to certain data, there will be another site that offers the same or similar data for free. This means that for any content provider willing to teach its users to pay for services, there is an almost impossible competitive barrier against all of those providers who provide the same or similar service for free.

One way to try to address the lack of direct payments from the users has been the advent of Internet advertising. For a while it looked like a promising

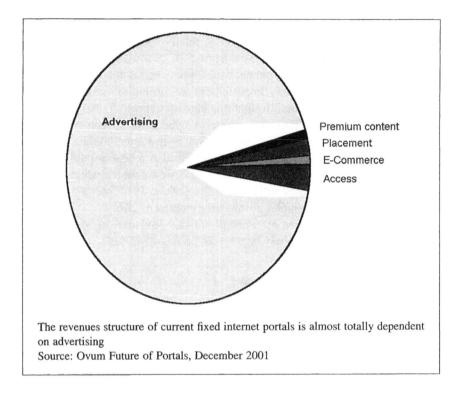

The revenues structure of current fixed internet portals is almost totally dependent on advertising
Source: Ovum Future of Portals, December 2001

vehicle to derive revenues to service and content providers, until the realities and mechanics of web banner ads and other Internet advertising became widely known. Falling click-through rates and lack of actual business generated by Internet advertising has burst the bubble that advertising alone can sustain an Internet business. A recent study by Ovum shows how strongly the Internet portal business is dependent on advertising revenues.

The irony of the fact is that for most regular users of the Internet there are several sites, often dozens, which are considered valuable and necessary. It is only that the users are reluctant to pay for access to these sites under subscription systems.

4.3.1 Small payments big problem

But even if the users were willing to pay, there is no easy *mechanism* to pay. Remember that the actual perceived value of most pages on most websites could only be estimated in pennies, often only fractions of a penny. Credit card companies will want a minimum payment, typically of the value of 5 dollars or more. Therefore, if the actual value of a given web page is a few pennies, the minimum payment destroys the ability to charge for the information. Of course, there are many ways that the Internet community has tried to get around this problem, with subscriptions, e-wallets, e-cash, etc. Nevertheless, currently there is nothing even nearly universal as a solution to collect something of the value of a few pennies on the fixed Internet.

4.4 Mobile Internet billing is exactly the opposite

The mobile Internet features the exact opposites of the fixed Internet. First, *all* mobile services have *always* been billable, and often perceived as quite expensive. There is no illusion in the minds of mobile phone users, that any service they might consume on the mobile phone could be – *or should be* – free. They know mobile services will almost always cost something. So there is no built-in resistance to paying for content on the mobile Internet because of history, of it 'always having been free'. The presumption favours billable services.

Secondly, there are no competing mobile operators, carriers, service providers or other entities attempting to provide similar services for free – *on the mobile networks*. While many fixed ISPs (Internet Service Providers) and other entities do attempt to provide free or very low cost services to compete with mobile services, none of those work for free *on a mobile phone*. So then, it means going back through the free fixed Internet model to provide competition. In the mobile/cellular world, there is no such thing as a free service. Even if the service is provided with no payment from the mobile phone owner, then there has to be someone else who is paying for the radio access costs, such as an advertiser or sponsor or employer. Every minute of airtime consumes a limited radio frequency resource, and thus every minute will be invoiced to someone.

4.4.1 Small payments, no problem!

Thirdly, and most importantly, the mobile operator has a built-in mechanism to handle tiny payments. The charging and billing engines that the mobile operators have, are massive in scale, totally dwarfing anything else seen anywhere else in banking, commerce, government and taxation, or the credit card industry. The UMTS Forum Report #16 explained the mobile operator billing system to good effect and its illustration serves to show the elements involved.

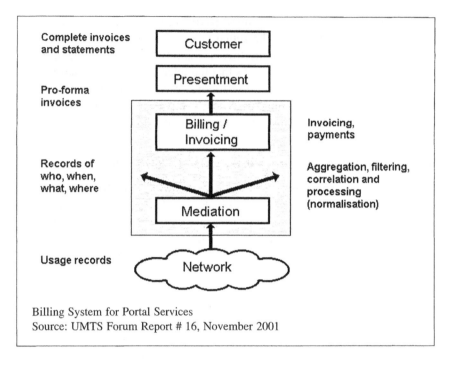

Billing System for Portal Services
Source: UMTS Forum Report # 16, November 2001

Imagine: a typical mobile network provider can tell you how long your local telephone call was, from your moving car to another moving car. The call might have only lasted 10 s. At the same time, the system identified you as the calling party, your network provider, the other party as the receiving party, their network provider, the location of placing the call, the location of receiving the call, the duration of the call and the charge, which you are billed. Both you and the called party could be visiting from another country and thus also be 'roaming' on other networks. This all for something that may cost 5 cents or less in actual consumed services. Actually, the network charging engine tracks much more even in today's second generation mobile systems. It knows which base station you were with when you started the call – and if you moved into another 'cell' in other words as your car moved, you might have moved into the coverage of another base station. The same for your counterpart. The system keeps track of which operator's networks the call went from, through and to. There may have been forwarding instructions – such as you having moved calls from your office number to your mobile number, etc., and the system of course knows your rate plans and how they affect the type of call you made, etc. There are systems which award discounts based on short number dialling, i.e. the system even keeps track of which buttons you pressed to reach the same recipient – and bill you a different tariff based on which keystroke set you used. The sheer amount of detail that is generated, analysed and stored for *every call* is enormous.

Nevertheless, cellular network operators do that for every call, for literally millions of subscribers, who place numerous calls per day. No other system in the world is massive enough to track such vast amounts of tidbit data of such small individual value. No paperclip manufacturer tracks sales by individual paperclip, no gasoline service station by ounce of gasoline, no bread bakery by slice of bread. All other industries consolidate sales into dozens, hundreds, or other larger groups of consumed items. Not cellular telecoms operators. In mobile telecoms, every second of airtime is meticulously tagged and logged and stored and billed. The actual heart of this is the charging system and the worldwide telecoms practice of collecting CDRs (Charge Detail Records). These form the basis not only of your billing information, but also of the telecoms interconnect accounting. As every phone can call any other phone anywhere on the globe, by definition the system has to be as robust everywhere for any telecoms operator.

4.4.2 Mobile Internet is where the money will be

It is currently very difficult to bill for services – especially for content with a tiny value, on the fixed Internet – but it is easy to bill for that content on the mobile Internet. This is a very compelling reason why content providers will be putting their best effort and their best content on to the mobile Internet. Of course, they can also put the content or some version of it on to the fixed Internet. However, if the money is at the mobile Internet, that is where the best content will soon migrate.

4.5 Customers and money will move content

When the majority of the customers are on the mobile Internet within a few years, and since it is much easier to bill for services on the mobile Internet, that is also where the best content will go. When you put yourself into the shoes of the content provider or application developer, in any industry from games to information to entertainment, etc., if you have the next stage of making improvements to your website and you have the options of fixed Internet and mobile Internet, the issue will become obvious, as the Americans say 'a No-Brainer'. On the fixed Internet there are less users, the users want free content, and it is difficult to bill. On the mobile Internet there are more users, the users are accustomed to paying, and it is natural to bill. Of course the best content goes first to the mobile Internet. Probably the content provider brings a variation or version of that content also to the fixed Internet. That may be simpler content or it may be a delayed delivery to the fixed Internet, not unlike how cartoonists use the fixed Internet as a delivery vehicle today. They put their new cartoon into newspapers which pay for their use, and then a few weeks later the cartoonist brings the cartoon to the free Internet site.

Table 4.1 Comparison of fixed Internet and mobile Internet

Aspect	Fixed Internet	Mobile Internet
Users 2003	Less	More
Cost of device	More	Less
Billing	Cumbersome	Easy
Micro-payments	Very difficult	Built-in
Willingness to pay	Want for free	Assume it will cost

This is what I meant with my vision that the mobile Internet will enable all content providers to be paid. As soon as the mobile Internet is the predominant content depository for new and updated information – and this will be sometime between 2003 and 2004 – then any content provider will have a means to set up the mobile Internet versions of their website and participate in the revenues generated by traffic to it. The actual syndicates and payment handling systems to enable this are still totally in their infancy, and will have to evolve a lot to enable the payments transfers. The Japanese I-Mode experience has clearly shown how this can turn a loss-making ISP into a profitable mobile and fixed Internet service provider, as the widely publicised pioneer case of Cybird has shown. And DoCoMo has reported long waiting lists to get on to its approved I-Mode (billable) sites as the demand by content providers has greatly exceeded DoCoMo's ability to connect all of the eager sites.

4.5.1 Migration will not be costly

Many reading this book will sigh and think that it is another dramatic transition requiring a million-dollar upgrade and vast new equipment, systems, specialists, projects, conversions, etc. For services which exist on the fixed Internet, to move to the mobile Internet involves literally only a fraction of the cost of setting up the fixed Internet service. This is because both are built upon IP technology and protocols. The big task was setting up the fixed Internet service in the first place. To port that on to a mobile Internet version involves mostly a redesign of the user interface of the data to fit a smaller screen size. All of the service logic, data storage, tracking, etc. systems can be reused in the mobile Internet version. A good example of how much easier it is to port a system than to create it, is from DHL the parcel delivery company, whose fixed Internet system to track sent packages took about 6 months, but to port it to its WAP version took 7 days. The migration will not be prevented by cost of migration, nor even significantly delayed by the time it takes to port the service. Most digital content projections suggest a fast migration and rapid growth in digital content, as for example in music.

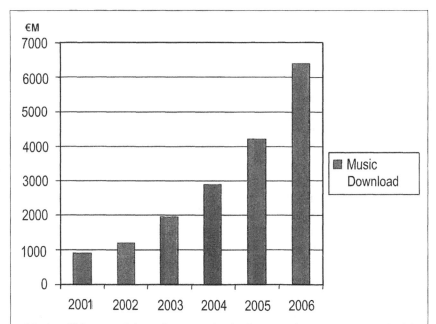

Music will be one of the major categories in the entertainment segment and will gradually capture a growing percentage of the total music sales market. Downloading music to new entertainment type UMTS devices will expand as the cost of data memory prices fall.

In some ways, the migration of content is good news also for all who use the fixed Internet today. Imagine all of those websites and content producers, putting their best effort into making the mobile sites *better* than their fixed sites. We all are frustrated by how many sites are poorly designed, with out-of-date information. The biggest reason why the content owners are not being more diligent is that they are not paid well enough to do it. The mobile Internet will soon be a good place for those who need to find information fast, but of course, it will be at a cost as all good things are.

4.5.2 Are not mutually exclusive

One has to understand that the various means to access the Internet, and the varying type of content that will be there, are not mutually exclusive and not a threat to each other. To use an analogy from another type of traffic, movement. We can always walk, and walking is free. We can get to almost anywhere by walking, as the comedian Steven Wright said ''Every place is walking distance if you have the time.'' But walking is very slow. Sometimes we would like to move a bit faster to get further in town, and we take a bus, and pay for our fare. But sometimes the bus does not go exactly where we want, or we are in a particular hurry, then we take a taxi.

Much in a similar way, there is a natural place for the free Internet. It will be there also in the future. Very much content will be there, free for all, and all kinds of people and companies and institutions around the world will work to placing that content there, and to maintain the sites with current information. Just like walking is free, we will always have a free Internet. But to find the exact information we want can be cumbersome and take a lot of time.

At other times, we are willing to pay a little for better content, or more immediate (faster) content. Much like taking a bus, we subscribe to a mobile Internet standard package of services and get access to better/faster information than what exists on the free Internet. We will be able to access the information from our mobile phone, but of course, that costs a little bit more than accessing the same information from a PC connected to a fixed line and the Internet. When we are in need of that kind of information, we are willing to pay a little bit for it.

Then there are those cases when we are in a hurry, need information really fast. Much like taking a taxi when we are in a hurry, we will access premium content sites, pages, and services on the mobile Internet. They will cost more than the standard content, but many of us will find those instances when we actually need the information so fast that it is worth the extra cost.

Just like walking has not eliminated busses, and busses have not stopped taxies from making money in all cities around the world, similarly the fixed Internet, the standard services on the mobile Internet, and the premium services on the mobile Internet will all coexist and have a market. As the majority of the users and almost all of the money are on the side of the mobile Internet, then eventually that part will have much more content than the free Internet.

4.6 Keep charges below the pain threshold

This chapter looked only at the micro-payments part of mobile services. Certainly it is not the only type of payments on mobile networks, nor are payments from users the only types of revenues by any means. Later in this book we will examine the way to build the money-generating elements into individual services in the Money Services chapter (Chapter 9), how to price services in the Tariffing chapter (Chapter 12), what part revenues play in the 3G operator business case in the Business Case chapter (Chapter 16), how revenues are shared in the Revenue Sharing and Partnership chapter (Chapter 15), and how money flows will evolve overall in the economy in the Money Migration chapter (Chapter 17).

This chapter looked only at the micro-payments, i.e. individual payments and charges worth less than a dollar. Even with those it is very important to understand the concept of pain threshold, and understand the economics of mobile operator subscriber numbers. Let me illustrate this by a simple example of an existing content provider on the fixed Internet.

4.6.1 Don't price at high end of what market will bear

The first example is the 'reasonable cost' approach, the one most content providers would most likely set if using their intuition and understanding of their customers. Let us assume you have a small but dedicated and loyal group of users on your website, for example a cooking website, and that user number would be 1000 regular users. If you were able to charge them 1 dollar per month to access your mobile Internet pages, and out of that traffic (since there always is airtime when using a mobile phone) for the sake of simplification let us assume you split the revenues 50/50 with the network operator, you would gain 50 cents per month for every person who accesses your mobile Internet site. So if you had 1000 loyal customers and one in 10 would be willing to use your mobile Internet site, you would gain 500 dollars per month, or 6000 dollars per year.

That is of course very much better than getting nothing from your fixed Internet site today. Even if you did get some advertising revenue today, that revenue from 1000 regular visitors would not be very much, and still all of the

4 Vignettes from a 3G Future: My Car, My Tour Guide

I really like my new car. It has the wonderful smart telecoms set-up. One of the neatest features is the voice combined with the car navigation. Whenever I visit a new town, I select the tourist guide which is tuned to me and my interests. It is quite surprising how much the mobile tourist guide can tell me about any town even when my interests are architecture, science fiction, and Formula One racing. Best of all, I can select the voice as well. First I had it speaking as James Bond, of course, and then been switching by my moods between a sexy woman's voice and that of Darth Vader.

While cars are likely to become ever more loaded with electronics and intelligence, one of the big factors is that users cannot access services via a keypad as they would on the handheld phone. Thus voice related interfaces are likely to develop very fast on car systems. Combining voice automation with preferred voice characteristics – the sound of your favourite actor/actress for example – with the words of a tourist guide, the control menu of the application, or even reading e-mail will enable 3G services to be used comfortably in the car.

900 who did not migrate to the mobile Internet version would still be active fixed Internet users, so most of your ad revenue stream would be intact. And most of the mobile Internet users would probably still *also* access the fixed Internet site, only probably less often.

The big problem with this approach is that it totally misses out on the mobile Internet population and its perceptions. At 1 dollar per month, a no-brand unknown website would have very little chance of succeeding to generate massive appeal. Your site would have a very hard time in attracting a huge following and their related large revenues.

4.6.2 Price as low as possible and gain mass market

Now let us take the mobile services approach. Let's hit below the pain threshold. If your website users were told that the mobile Internet version costs a dollar per month, many would not want to use it. But if it cost 10 cents to use, *most* would feel it is such a trivial additional cost, they would not mind. At 10 cents per month – meaning that the operator gets 5 cents – there cannot be very much traffic load per user per month. I am now assuming that your website is a typical Internet website with static information, not any streaming data or video clips, etc. But at 10 cents per month, most of your current users would probably migrate. Let us assume it is only half of your users, or 500 users. From those users you get 'only' 50 dollars per month or 600 dollars per year.

However, now your website costs only 10 cents to use for *any* mobile Internet user. At 10 cents it is very likely to be well below the pain threshold of most people interested in cooking. Let us assume that you get listed in search engines for the mobile Internet. And let us assume that one person in 20 is interested in cooking, and that out of those, one in 10 is interested in looking for a new website. Furthermore, let us assume that your website is that good, that if a cooking-interested browsing person lands on it, there is a one in five chance of them remaining loyal. Let us plug these into the magic of mobile services. In the UK alone, the four mobile network operators have subscriber bases of about 10 million per operator. Let's start with your service introduced on one of them. If one subscriber in 20 is interested in cooking, that results in 500 000 potentials on that network in the UK. Out of those one in 10 visits your page once (50 000) and out of those, one in five remains (10 000). When you gain 10 000 users paying 10 cents per month, your website generates 1000 dollars per month, after the 50/50 split with the operator, it results in 500 dollars to you per month, or 6000 dollars per year.

Out of ONE network operator in the UK, you have now a revenue stream of 6000 dollars per year from new users, as much as you would get if you squeezed the maximum out of your existing customer base. When you add your below-the-pain-threshold pricing revenues from your current customers, your revenues are already more than in the first example, i.e. 6600 dollars per month.

Now let us add the other three network operators in the UK. Of course you could bring your page to the other three operators and now you have four times 6000 dollars or 24 000 dollars per year, plus the 600 dollars per year from your existing customers. And that is only the UK. On an English-speaking service, you could automatically port it to all other English-speaking countries like the US, Canada, Australia, New Zealand, Ireland, South Africa, Hong Kong, etc. With the early revenues you should be able to invest in an automated translation system to port your mobile Internet website to Spanish, Chinese, Japanese, French, etc. languages and cover most of the world. Charging 1 dollar per month would bring you a trivial income stream. Charging 10 cents per month could bring revenues in the six figures within the first year.

The key is to keep the pricing below the pain threshold. Then the massive subscriber numbers will bring in the revenues. Even if you only get a *penny* per subscriber per month, and if your service is appealing to only one person per hundred – if you were to reach the whole global mobile subscriber base, your *monthly* revenues would be 100 000 dollars, and you would make a million dollars in less than a year. This at your revenue of 1 penny per month. You do not need to price near a dollar, the lower the better. A good benchmark is the current pricing of one SMS (Short Message Service) text message (between 10 and 15 cents per SMS depending on country). Keep your pricing below SMS and it will be below the pain threshold for most services.

4.7 Money money money

When we think back to the Internet before web browsers, the users were researchers and students. The content was free. There was no business. There was very little entertainment. It was not a mass-market media tool. The web browser changed that. In addition, the PC penetration made the original population an insignificant minority on the Internet. Currently the Internet has information and entertainment, a lot of advertising, and commerce. Individual pages of content have value, but the system is not naturally able to handle the billing. Moreover, the current population is accustomed to content being mostly free.

The near future expands the Internet-connected population to where the dominant device will be the mobile handset. The people using mobile devices will feel it natural to be billed for using the mobile device for any service. The content provider will bring the best content to the dominant access device.

The PC-based fixed Internet will not vanish, it will still be there, but the mobile Internet will have the best content. The evolution towards the mobile Internet with micro-payments and best content seems inevitable. Yet, this is still only an analysis of existing trends, and a projection of them into the future. Still, if micro-payments enable content providers to be paid, we may have to change Robert Graves's famous quotation: ''There is no money in poetry, but then there is no poetry in money either.''

5

'A prince who will not undergo the difficulty of understanding must undergo the danger of trusting.'
Lord Halifax

The 5 M's of 3G Services:

Recipe for killer cocktails

What makes a mobile service different from other services? How can a 3G service be built with more value so that the user is willing to pay for the service? And how can services in the mobile Internet be built so that the operator and content provider can also make a profit. Many of the attributes have been discussed in previous chapters, and some methods such as PAIR, MAGIC and 0-1-2-3 have been discussed there. But these all have their shortcomings.

There are actually five key elements of 3G services that allow for value to the user, and profit to the operator. By considering any potential service through these five dimensions, optimal service definition can be achieved. The dimensions are called the 5 M's of 3G Services. I have been developing this theory with Joe Barrett of Nokia, and we introduced the theory in our book, *Services for UMTS*. The theory covers all aspects of how to create end-user value that can be turned into profit. Therefore, the theory of 5 M's is a good guide to discussing services from the profit point of view. The 5 M's is not specific only to 3G networks, and the same 5 M's can be used to create value to any mobile services on any mobile network technologies.

5.1 The 5 M's

There are five significant and distinct attributes for mobile services, which define usability, utility for the user, and which define cost effectiveness and invoiceability for the network operator. The five attributes are called the 5 M's. The 5 M's are:

Movement – escaping the fixed place
Moment – expanding the concept of time
Me – extending myself and my community
Money – expending financial resources
Machines – empowering gadgets

Any service that brings value through any one of the 5 M's has potential for success as a mobile service. In most cases services naturally address a few of the five, and the more of the 5 M's that are included, the more the service becomes particularly suited for the 3G environment. In most cases services can be enhanced by adding to any of the five dimensions, and the most appealing applications – and any 'killer applications' – will use most if not all of the five.

5.2 Movement – escaping the fixed place

Movement as a dimension of a mobile service is probably the most obvious when considering mobile phones and the networks that support them. The Movement attribute covers all of the significant aspects of the *possibility* for a service or user or device to move. It includes of course mobility – meaning that the mobile phone and user move about whether by walking or driving a car or travelling in a train – but includes also concepts such as locality, globality and home base. Movement as an attribute makes it possible that the mobile service will be just right regardless of location, regardless of motion and appropriate for the occasion.

Movement: Mobility, Locality, Globality, Home Base, Positioning

Mobility means that the service transfers with you as you move. The most obvious service is for automobile drivers, looking for the nearest hotel, etc. The service should keep track of where you are and dynamically offer you the most appropriate service or information as you move, such as the nearest hotels in that town or neighbourhood. The service should move beyond any location and its information automatically as you drive on past it.

Locality means that the service provides specific locational information. If you are visiting another city and ask for the weather forecast, the service should provide local weather, not that of your home-town.

Globality means that some services need to provide the same service worldwide. A good example of a Globality service would be Wall Street financial news.

The home-based service means that it knows where your home base interest is, and provides automatically, or upon easy request, the home-town information. If you are at the airport in Spain, returning home, and ask the service what will be on TV, you would want access to your home country TV listings – where you usually are based – not those of Spain which you are just about to leave. A technical solution to this is called Virtual Home Environment, or VHE, which is described later in the Movement chapter (Chapter 6).

Positioning is identifying the physical presence of the mobile phone (and its user) or other 3G device. Positioning rarely provides direct benefit to the user, unless the person is totally lost and wants to know where he is. Nevertheless, positioning can be used to guide and to provide assistance in finding friends and colleagues for example in a crowded place.

Movement services need to be intelligent in their handling of this location information. If you are visiting another country, and looking for a cash machine, it should offer you the locations in that city, not your home city. If you want to get a stock market update, you probably want your own country's stock market, not the local stock market of the country you are visiting. In the case of TV listings, you probably want local listings, but if you fly back home today, or want to activate your VCR remotely, you want easy access to your home-town TV listings as well. Therefore, the Movement attribute services need to behave intelligently for the user.

5.2.1 Movement is unique to mobile networks

Movement as an attribute is the most defining aspect from mobile/cellular networks, and what sets them apart from other networks. From Movement the 3G operator can find competitive advantages that other technologies cannot easily match. With concepts such as roaming and remote access, this is the part of the 5 M's that the current mobile network operators know the best.

5.3 Moment – expanding the concept of time

Services that use the Moment attribute allow for postponing, re-scheduling, and last minute behaviour, as well as catching up on lost time. Busy people use their mobile devices often to make last minute modifications to plans, or to leave a decision to the last moment. Technology such as voice mail, e-mail, SMS (Short Message Service), etc., enables us to receive messages when it is convenient, thus effectively moving time or actually by moving communication to another point in the future. Manipulating time also sometimes works backwards, for example having easy access to yesterday's news.

Moment: plan, postpone, fill time, catch up, multitask, real time

Planning is critical to efficient use of time. Planning can be greatly enhanced by a device, which is with us at all times and connected to the rest of the world. A good example of a planning service is a network scheduler where you can
see what your team members are doing to try to find a common time for a meeting.

Postponing includes making last minute changes and last minute decisions. A good example is waiting for some other events to finalise before settling on a specific time to meet friends; that we all have mobile phones allows us to do just that.

Filling time involves doing things with sudden idle time, such as catching up on e-mail while waiting at an airport, or reading the sports updates while in a taxi.

Catching up means trying to 'recover lost time' or more precisely to do something for which there was not time when it was initially intended. An obvious example is reading yesterday's newspaper.

Multitasking is the ability to expand the concept of time and includes functionalities to do multiple things at the same time. They do not need to be all done on the mobile phone. A good example is that the mobile phone allowed us to talk while we walk, and thus we can get somewhere while we conduct business on the phone, hence multitask. The same concept works in the car; we drive and talk at the same time.

Real Time is delivering services to you when they actually occur. The typical example is stock prices, where half an hour can easily mean the difference between a profit and a loss.

5.3.1 Moment activates

Moment is the most activating attribute of the 5 M's. If the need is exceptional, and the user is already aware of the availability of a given service, the Moment attribute can result in almost compulsive action by the user. For example if you are in an elevator and you hear two businessmen talk excitedly about news about your competitor, you would like immediate access to that news. If you can act upon it then and there, you probably would be willing to pay a considerable premium – meaning a dollar or two in value – to see the news. But if you cannot access the news until you get back to your office at the end of the day, you can then go to the Internet and read the news for free. Hence, it is very important to understand the Moment attribute, and to create as much opportunity as possible for the user to fulfil momentary – often temporary – needs and wants. The far-sighted operator will train its users to use momentary features to create almost addictive uses of the Moment services.

5.4 Me – extending me and my community

The Me attribute is the most powerful of the 5 M's. Me includes the personalisation of the mobile service, and its customisation of services to its users, as well as the extension of the self into the various communities where the person has a presence. The Me attribute is why we view the mobile phone as so much more than just another personal electronic device. As personalisation evolves, so too will the Me attribute's importance keep on increasing as the innermost measure of whether any given service is truly for us or not.

Me: customised, relevant, community, permission, presence, multi-session

Customised means personalisation of aspects such as the language we prefer. The language choice is not automatic by country of birth and the 'mother tongue' – as has been seen in many cases where a person may set the mobile phone interface to a second or third language just as practice on using that language. Customised ranges from the very 'serious' such as selecting personal service bundles from the available service portfolios, to the 'frivolous' such as colours, screen savers and sounds that our phone and services will use.

Relevant means providing those aspects of a broad service, which actually have meaning to the user. A typical example is the horoscope, while a newspaper prints horoscopes for all 12 signs of the zodiac, the mobile service does

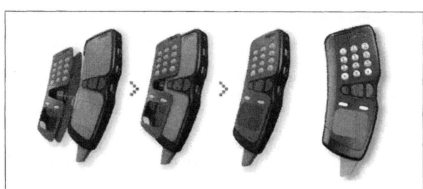

There are many variations of new phone designs. Gitwit have one interesting view with slide on covers and the screen and antenna at the bottom of the phone

not have to show all, usually only the user's own horoscope – and perhaps a few loved ones if one is into horoscopes – is enough. Rather than showing all sports results, only show the basketball scores, and so forth.

Community means the ability to interconnect and share with the various mostly informal communities that one is involved with, which could be family, work colleagues, friends, people sharing the same hobby, etc. For example, a service to help coordinate the families in taking the boys to football practice.

Permission is the control of who has access, and in what ways and at what times to your phone. A simple example is setting up your phone to ring in a different way when your wife calls you.

Presence is the ability for the network to determine status of a given phone or device and allow services to be tailored to that condition. For example using the 'I am not at work setting' if it is past five in the afternoon, and a work colleague calls at that time, forward the call to the office voice mail.

Multi-session allows multiple concurrent connections on one phone. A typical example is to view an advertisement for a movie and at the same time talk on the same phone with someone to decide if you want to go see it.

5.4.1 Me binds

First, one cannot overestimate the importance of the Me attribute. You might feel that the ability to customise the colours or sounds of a phone is totally meaningless, yet millions of young people make phone purchase decisions almost exclusively on how 'cool' they look and sound. That is why the Me attribute is so powerful, what may be meaningless to you may be the most important feature for the person next to you. This point may need a lot of re-emphasis especially with telecoms operators, equipment vendors and IT companies, as these companies tend to be very engineering-oriented. While *you* may view your mobile phone as a utility device, and consider it primarily for its functionality to your work – remember that most people – and the mass market for mobile phone fanatics is the under 25 age group – are not like that. They really do want their 'frivolous' aspects and *will* make decisions based upon them. If you work for a telecoms operator (carrier), equipment vendor or IT company, never make the mistake of assuming the mobile phone mass market is people like those you work with; they are not, and you will not find the mass market until you acknowledge that.

The Me attribute is the biggest factor in keeping users on a given service or network (or loyal to a given mobile phone handset). It is the most powerful of the 5 M's, and its importance cannot be overstated. However, in most cases the Me attribute is the one which network operators know the least well. In fact the Me attribute is often dismissed and even actively ignored. Readers of this book should keep in mind that the Me attribute is the most powerful of the 5 M's, ignore it at your own peril.

5.4.2 Me builds communities

It is also important to understand that the Me attribute is an extension of myself into my communities. At work, a group of colleagues who work on the same project could be such a community. They can communicate effectively via a number of devices, fixed and mobile to achieve better productivity. If the members develop their community further, they might go out for some beers after the launch of the product, or after reaching some important project milestone. In Finland it is very common that when a new team is set up, one of the very first activities is that everybody enters the mobile phone numbers of everybody else into their mobile phones, as a first step to being able to be directly connected. It results often in a group 'ping-pong' game of SMS traffic, everybody sending one test SMS to each other, to test and verify they have the right numbers.

The 3G mobile terminal offers numerous ways to build and maintain communities further by allowing communication, contact, caring and sharing between members. Note that while the Me attribute binds by itself, built-up communities bind as well, further enhancing the way that Me binds the user to the network operator. In other words, if we have our community set up with one operator, and all of my community friends know my number there, I have more reason to remain with my current operator and not change numbers, etc.

The Me attribute is that attribute which the individual user will value highest. It is a reflection extension of the self, the ego, the person, as well as the person's connection to his community and a reflection of his image in it. The operator should build a strong 'I understand you' feeling to its services, honestly listen to customers, and deploy service platforms, which can respond to individual personal needs.

5.5 Money – expending financial resources

The ability to spend money or use the mobile terminal for money transactions is another useful aspect of services for 3G. The mobile phone can also be the electronic means of delivery for a service, such as a song in digital format, as well as a means for advertising it. The network connection allows for easy charging and billing of the service that is difficult on most other networks. Thus, the Money attribute is a key to bringing high quality content to 3G.

> Money: m-commerce, micro-payments, m-banking, m-wallet, m-advertising, spon-sorship

m-Commerce is a term that is still evolving. Some count any commerce transaction, which has included a mobile component to be m-commerce, i.e. if I see an ad of a tennis racquet on a mobile Internet page and then walk to the store and buy it by paying cash, it would still be m-commerce by this defini-tion, as I did use my mobile phone in the decision-making process. Others define it very narrowly, by which the purchase must be completed on the mobile terminal, e.g. I purchase a Rolling Stones' CD on my mobile phone and download the songs on to my phone, only then is it m-commerce. For the purposes of this chapter, it does not matter much how broad is the definition, rather the point is that for the 3G network operator and content and service providers, they need to consider all opportunities to engage their subscribers and customers into m-commerce. Of course, the more of the transaction that can be completed on the mobile terminal, the more relevant it is as a mobile commerce service.

Micro-payments are commerce transaction payments whose value is less than 1 dollar and were discussed in its own chapter earlier (Chapter 4). A typical micro-payment purchase is buying a can of Coca-Cola from a vending machine.

5 Vignettes from a 3G Future: Competitor Update

One of the best services we have set up at our company is the mobile version of the competitor update. This service is priceless. Every day our competitor intelligence team selects the top three or four main stories and sends them to those of us who have subscribed to the service. We get the brief summaries directly to the phone, with links to see the longer story if needed. The sales force cannot imagine life without this.

An operator can create a tailored intranet portal access solution for businesses from very small to corporate. The smaller the business, the more likely that it would want to outsource the whole solution. The operator can tailor the services so that the intranet access has various controls and safeguards to protect sensitive data. The system could integrate various work efficiency tools such as schedulers, mailing lists, internal database links, etc. A business personalised portal service would be particularly 'sticky' in that it would tie into the whole way of working for a large portion of the employees for the operator, and thus ensure they would not want the service interrupted or moved.

m-Banking is interactive banking on the mobile phone. A typical service would be paying your electricity bill by using the mobile phone.

m-Wallet is moving cash and credit cards from the wallet to exist on the mobile handset. The m-wallet would for example have your credit card information and rather than giving a merchant your plastic credit card, you might just point the mobile phone at a device and authorise the payment by clicking on the keypad.

Mobile advertising is providing advertisements via the mobile phone. A typical example is a mobile coupon that is sent to your phone and you can redeem a discount by showing the advertisement at a shop.

Sponsorship allows a third party to provide free (or reduced price) content. An example is the tourist map of the city sponsored by McDonald's, and the map would conveniently display the Golden M signs of the restaurant chain at their various locations.

5.5.1 Money brings content

The Money attribute is the one valued most by the content producers and service partners. It is the magic key to revenues for services, especially where the individual value can be measured in sums as small as pennies. The 3G operator should enable all types of Money features and help make any transactions possible. Here the direction is very much towards the various merchants and all content providers, by providing them with an outlet, revenue stream, and billing system. The services can be built with options and features to provide transaction histories and billing information, etc., as parts of value-adds to consumers and as billable services to partners and content providers. The micro-payments option is the most dramatic single element of the 3G environment, and it will enable the migration of content from the fixed Internet into the 3G environment. The Money attribute must be considered for every service, but many service creation engineers and product managers and marketing managers are unaware of the various aspects of the Money attribute and may be ignoring vast additional revenue possibilities.

5.6 Machines – empowering devices and gadgets

The last major area of beneficial services is allowing machines to perform activities and communicate. They can communicate with each other such as your car noticing it needs an oil change and scheduling the service visit with the garage. The machines can communicate with us such as sending us an alert when the stock price hits the specific price. Moreover, the machines can accept our communications such as automated ticketing systems for airline seats. This type of listing will probably never be long enough, so just some early obvious areas are automobile telematics, home appliances, metering devices both fixed and mobile, robotics, voice activated automation services, etc.

> Machines: telematics, metering, remote access, appliances, robotics, automation, connecting with just about anything

Telematics are machine-to-machine communications. The most common example is the car communicating automatically with its surroundings, such as checking on traffic congestion and suggesting a route with less traffic.

Metering is the remote reading of various meters and sensors, such as a billing system remotely reading the gas or electricity meter.

Remote access is the ability to connect with controls of systems and make adjustments. A good example is remotely accessing your home to turn on the lights or heat.

Appliances are the myriad array of devices, gadgets and appliances and any utility that can be built into controlling them from afar. For example, you might want to access your VCR suddenly when you notice that you are running late and you are about to miss your favourite TV programme.

Robotics could fall into the categories of remote access or appliances, but they are listed separately here as a reminder that the connected machines may be mobile devices themselves. The access may include vision or sound, e.g. a TV camera on a household robot controlled by a 3G phone.

Automation refers to most other automated systems, not necessarily physical devices. Automation would include computer programmes and systems. The mobile service could access an accounting system and make entries of a payment having been received.

Connecting just about anything, anywhere, is the final catch-all term, as a reminder that the above examples are only the tip of the iceberg. There are hundreds of thousands of different possible connections that could be made in the area of Machines.

The Machines attribute will connect and enable a population of 3G network users that will dwarf the human population. Early estimates suggest connected devices will soon have twice the population of humans, and later in the 3G network life cycle the amount of non-human users might be 10 times that of people. Some of the machines will produce only small amounts of traffic, but others might produce more traffic than the most productive humans. The Machines aspect is the biggest potential area for growth in 3G services.

5.6.1 Machines reduce costs

The Machines attribute is the key to stronger profitability, as the ability to provide automated services through gadgets and devices and automated servers is a significant cost-saving dimension to 3G services. Numerous automated systems already exist in telecoms, such as the keypad controlled menu systems to guide you to the correct department in an automated voice response system, to cases where you talk to a machine and it replies in voice. In addition, where machines can 'talk' to other machines to build and tailor the service as requested, the connectedness and machine-to-machine dialogue extends that cost efficiency down the delivery system of the service logic.

The 5 M's

Movement – escaping place (local, global, home base, mobile, position)

Moment – expanding time (plan, postpone, stretch, fill, catch up, multitask, real time)

Me – extending myself and my community (personal, relevant, customised, community, permission, language, multi-session)

Money – expending financial resources (m-commerce, micro-payments, m-banking, m-wallet, m-advertising, sponsorship)

Machines – empowering devices (telematics, machines, appliances, robotics, auto mation, connecting with …)

5.7 Using the 5 M's to build value to a service

When designing a new service, the operator or service provider should try to include as many of the 5 M's as possible. In most cases the key benefits of a given service will involve a few of the five, but often several of the 5 M's can be enhanced by building the service even more valuable. The 5 M's can be used to enhance services further, in response to competition, as each of the 5 M's is a dimension, and you can move further into that dimension to build added value.

To illustrate how the 5 M's can transform a service which exists already, by way of example, let us take a typical digital service, which exists today on the free fixed Internet. By using the 5 M's the service can be enhanced to create more value to the user, and that the user will be willing to use the mobile version of the service.

5.7.1 Train timetable and the 5 M's

Let us assume that we have the train timetable for the local trains in a big city. The timetable is printed in paper format and is also published on the fixed and free Internet. The Internet version of the service includes the ability to search by train route names, train stops, and so forth.

5.7.2 Info is already in digital format

The Internet service is already digital, so it contains most of the information that the service provider would need to provide to create the mobile service. The additional parts needed to the fixed Internet version would be real-time (Moment) information of the actual location (Movement) of the trains. These are actually already also in use in many cities where the train system reports how soon the next train will arrive at a given train stop. So even this information is often already being collected and communicated (and exists in digital form).

5.7.3 Enhance with Me attribute

To start with, let us take the Me attribute. Of course, any person who would want to access a train timetable might be interested potentially in any train. But much more likely, if it is a weekday and the person is of working age, the person would like to know about the train(s) that go between home and work. Since the 3G network provider knows the billing address of the phone user, the network provider knows (in most cases) the home address of the mobile phone user.

The mobile Internet train timetable service could offer, as the first choice, the schedules of those trains that go by the mobile phone user's home. Of course there should be links to all schedules, but odds are that the 'home trains' – those going to, or leaving from the station nearest to the home – are most often of interest. The service could easily be built so that the user could make selections of what routes or destinations interest the user, and those would then show up on the 'shortcut' schedules. As services become more advanced, the system could keep track of what trains are of interest, and remember those close to the top of the list.

5.7.4 Add Movement and Moment

The movement of the trains is already collected by the system, so it only needs to be integrated with the mobile service. The same is true of course of time (i.e. the network clock keeps track of time). So the system could now offer me as my first choice, the *next* train (Moment) to come to the train station where I *happen to be standing* (Movement) and which goes near to *my home* (Me). There would be of course the chance to scroll and search for other train timetables in case I happened to need it. Nevertheless, already now, the mobile service is much better than what is available on the fixed Internet.

5.7.5 Add Money and Machines

Someone has to pay for using the service. It could be provided on a query charge, for example 1 cent per query. The content is digital text, no images, so there is little to transmit, and the cost could be low. But better yet, if the service could be sponsored – for example by one of the afternoon newspapers, showing the headlines of today's paper – then the service would be of even better value to the user. Finally, by automating the system so that no human is needed to update the information, and offering a push link service based on the train station location, the service is complete. The push link would be sent whenever I am at the train station. Behind the link would be the next train information as well as the sponsoring newspaper headlines. There would not be a charge for receiving the link, but if I clicked on the link to view the timetable (and possible sponsored headlines), a charge would be formed. A fully automatic service would learn what I like to do, and start to anticipate what types of services I might request next. A natural extension of the service is the ability to pay for the train ticket directly on the phone so that I don't have to stand in line at a ticket counter nor to carry cash for the train ticket.

Now we have created more value into a mobile service out of content, which already existed, in digital form. The mobile version of the train schedule is much better for the busy commuter than the current version on the fixed Internet. All five of the 5 M's were used. Any service can be enhanced with at least some of the 5 M's. The more of the 5 M's that can be included, the more the service becomes beneficial.

The 5 M's are each a continuum which can be extended further and can be thought of as an arrow expanding from zero. The user will not particularly care about the absolute value of the attribute, but will be very aware of the *relative value* in contrast to similar services offered by other providers. Therefore, it is very important to keep abreast of what the competitors are providing and plotting the relative merits on a 5 M's chart. Then adding features and ability to your own service so that the user perceives better performance on at least some of the five attributes.

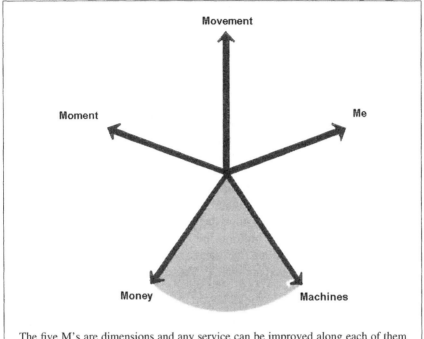

The five M's are dimensions and any service can be improved along each of them

5.8 The killer app in 3G

I feel compelled to discuss the issue of a 'killer application' for 3G some-where in this book, and decided to address it here. The 'Killer App' is no James Bond-like stealthy assassin's tool to use the phone to murder someone. Killer application means the application (or service) that alone provides a compelling reason to purchase a device.

For the original PC the primary killer app was word processing, although the spreadsheet was a close second. I bought my first PC in 1990 because it was the only way I could run my favourite spreadsheet programme, Lotus 1-2-3. Spreadsheets were therefore the killer app for me when buying my PC. In fact, Lotus 1-2-3 was the reason also why I purchased my first PDA – an HP 200 LX which had Lotus built in, in 1994. For the early Macintosh the killer app was desktop publishing. For GSM (Global System for Mobile Commu-nications) mobile phones, the killer app was voice, but recently for the younger generation, the killer app is becoming SMS text messaging.

So, what will be the killer app in 3G? There will not be one because 3G will not be a single-use device, but rather it will have hundreds of services/applications. To look for a killer app in 3G would be like looking for one on TV. Maybe one could say that for radio in the 1940s and 1950s there were killer apps, music and news probably. But by the time TV came along there were too many different types of programmes catering to different audience segments, that a single killer app could not be identified. Most people consume numerous types of programmes on TV. What was *your* reason to buy your first TV? Some are news junkies; have to have their news. Others are sports fanatics, and the TV was bought to see football, or car racing, or baseball, basketball, tennis, you name it. Others need their soaps – soap operas. Others cannot live without talk shows. Others want game shows. Others want drama, movies, comedies, cartoons, etc. etc. etc. Often it is not a single programme on the air now, but rather the notion that the box will entertain and inform us. In addition, in very many cases, we have many different services that we want, and their *combination* is our personal killer app. There is no single killer app on TV. There will not be a single killer app on 3G either.

For 3G there will be compelling reasons to buy, which will vary by individual. Notice also, that in most Western countries those who will buy a 3G phone already have a 2G phone today. So for most, it will be an upgrade decision, not an initial purchase. So while voice/telephone ability and SMS may be the killer apps for 2G, in 3G they are taken for granted, and one will need a compelling reason to *upgrade*. That for some will be streaming services like music and video. For others it will be access to information. For others it will be multimedia capacity in messaging. For others it will be that their favourite game can only be accessed on that device. Yet, for others it will be the community services, and so forth. There is no single killer application in 3G.

It is very important to note the difference between killer application and profit. The killer application often is not by itself a major source of revenue, nor the key to profits. It is the *activating agent* to make a purchase of *something else*. When we bought a PC, whether the killer application was the word processor in the 1980s or e-mail and the Internet in the 1990s, the software was not the major cost of our purchase, in fact some of the software was distributed free of cost. We bought an expensive personal computer costing about 2000 dollars and often a lot of other accessories to go with it. The killer application was the reason to buy the computer, not the most profitable part nor the one generating the most revenues.

Furthermore, that there is no identified killer application in 3G should not worry the industry. All of the major elements of the early 3G business case have been validated during 2001 – the viability of the business in voice calls, messaging traffic, data networks access, mobile commerce, mobile advertising and telematics traffic – so a service offering all of them is on very sound business basis. A 'killer cocktail' is very viable and most people will find numerous appealing services from just those listed in this book. While video calls and music streaming are still unproven, their contribution to the business case of 3G is so miniscule, that even if they fail totally, their impact can be lost in a rounding error. So while it is fashionable to look for a single killer application in 3G, it seems rather obvious that one single application will not emerge, and that the business can be on very sound basis without one. We all have our own personal killer cocktails, and will make our upgrade (or initial) 3G purchases based upon our own wants and desires.

5.9 Finally on the 5 M's

The 5 M's are one way to identify the attributes for successful services for 3G. The 5 M's help guide in designing services that are desirable, valuable and 'sticky'. The relative merits of any of the five attributes are relative to competitive offerings in the marketplace, both on other 3G services and via other means. Five M's are useful in understanding which services might be popular, and how to make any given service *more attractive* in the 3G environment. Note that the 5 M's are not a system for categorising services as any successful 3G service is likely to use most or all of the 5 M's. We offered a system for categorising mobile services in the book *Services for UMTS*, and if you would like to understand that more, please see that book. For understanding the profitability of mobile services, it is not really relevant how any given service provider actually groups services.

For success in 3G, the 3G operator and its partners must deploy good services, which will bring value and utility to the user. The 3G operators will be providing those services in a heated competitive environment where many of the world's biggest corporations are fighting for their slices of this new cake. There will be winners and there will be losers. An ideal service would be one, which is very fast to spread within its own network, and one, which the competition is not quickly able to copy. The ideal service should be priced low enough to get massive adoption, but still high enough to bring solid profits as long as it is an exclusive service for the 3G operator. That requires a lot of creativity, even with tools like the 5 M's. And at the end, a good dose of humour will help ease the stress and help things along. While I believe fervently in the theory of 5 M's, and one might say it forms the very philosophy of my thinking on new mobile services, there are doubtless going to be other competing philosophies. For those I humbly offer this guidance: 'Utinam logica falsa tuam philosophiam totam suffodiant!' (May faulty logic undermine your entire philosophy!)

6

'The journey of a thousand miles starts with a single step.'

Chinese proverb

The Profits of Movement Services:

Escaping the fixed place

Movement, the first of the 5 M's of 3G services, is the attribute, which is unique to mobile networks. Movement is the attribute easiest to associate with mobile phones and services on the move and Movement is also the feature that mobile operators know best. Mobile operators can fairly claim a significant portion of the revenues and profits of any Movement-oriented services as these could not easily be created on other technologies. There are several aspects to Movement, such as locality, globality, home base, and mobility which were discussed in the previous chapter on 5 M's. Movement means the ability for the service to escape limits of location, and meaning that the service will seem perfect regardless of location. The Movement attribute is a natural competitive advantage for the mobile operators when compared with fixed operators, and thus any Movement benefit should be able to be monetised to a large degree.

When considering services in this and the following chapters, please keep in mind that typically services benefit from including as many of the 5 M's as possible. The other 5 M's are Moment, Me, Money and Machines. Thus, most services in this chapter could just as well be listed in one of the following chapters, and many of the services in the following chapters could be listed here. There is a lot of overlap among the 5 M's in a good 3G service. There are hundreds of services that have a high benefit on the Movement attribute, and a chapter like this can only explore a few of them. The next four chapters will examine the other 5 M's.

This chapter will discuss some of the profits in services, which have a strong benefit from the Movement aspect of a 3G service. The services are not in any order of importance and this brief discussion will not be able to adequately address even the major areas of Movement type services. However, a deeper discussion of a few services is useful to understand how the revenues and profits are derived from the Movement aspect of 3G services.

6.1 Guiding services

The best example of guiding services is the 'how can I get there' service. The network keeps track of where the mobile terminal is at any given point in time. Whenever the owner wants to find his/her way to any other location, he/she can simply ask the mobile phone to give guidance. In its simplest form, the guiding service could offer a presentation of a map and show the location of where the person is currently, and where the destination is.

But even simple improvements can be made to this service. Some people are naturally very comfortable reading maps, while others are very uncomfortable trying to interpret maps. By spelling out the instructions into words, the service can be made to be much more useful as a guidance service especially to all who are not natural map-readers. For example, the guidance can say 'turn left at the next crossing to Smith Street'. Instructions like these can then be read by a text-to-voice conversion programme, which can read the instructions out loud, as if one had a personal tour guide helping with the journey to the destination.

Guidance advice is something we often need, but can rarely predict that need. We never *plan* to become lost. Thus, the 3G operator should offer a very basic tourist guide as a standard feature of entering the network for all roaming travellers who appear in the network. The value of the services is dependent partly on the maps and location information, where the value of the service would be enhanced by the location information and where needed, the map generating and scaling systems would bring part of the revenue to the mapping software.

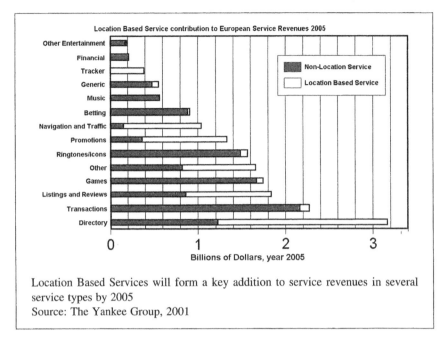

Location Based Services will form a key addition to service revenues in several service types by 2005
Source: The Yankee Group, 2001

But for the actual content – where that is something beyond only a map – for example listings of Italian restaurants or guides to the local theatre and sporting events – these would depend on creating and maintaining the content.

6.1.1 3D virtual city

A more complex solution can be made with virtual models of the town in question. Many towns and cities have virtual 3D (three-dimensional) environments created which replicate the real buildings in a 3D appearance. These can give a visual imagery of the route to be taken, which looks like one had video taped the actual journey while walking the route. All buildings seem the right shapes, colours. They 'grow' as one approaches them, all in real 3D appearance. The trip to be taken can be shown as a 3D journey, helping further in finding the destination. The scenery of alternate routes can be evaluated by viewing alternate routes. Again, many who have seen 3D presentations prefer them to plain 2D maps.

A final dimension to the tour guide is the sponsored tour. The tour guide service can also be a sponsored tour, where out of alternate possible routes, the offered route would include a slight detour, and go by the sponsoring company, such as Pizza Hut or McDonald's. The sponsoring/advertising does not need to be blatant and overpowering. Just seeing the restaurant logo along the way should be enough to draw the hungry traveller into that fast food restaurant rather than its competitor. For those who are simply not hungry, no amount of aggressive advertising would be enough to make them buy food.

> **Hint for mobile operators: provide more than available otherwise**. Don't just replicate what is available in the market, make your service better in some way so it is unique.

6.2 Adding value to travelling life

Another major area using the Movement attribute is travel-related services. These are very easily recognised from Internet- and paper-based tourism, business travel, and frequent flier programmes. Typically, travellers need to find hotels, restaurants, taxis, cash, etc. They want to know currency rates, exchange rates, time differences. Travellers have personal interests such as wanting to visit museums, catch the cultural events, visit nightspots or go shopping. Travellers often have sudden changes of plans and services to help last minute bookings, changes, updates, etc. can be very useful. A separate category are services related to travelling by automobile, which will be discussed in depth in the Machines chapter later in this book (Chapter 10).

6.2.1 Connecting travel

Travellers usually have a need to use multiple means to complete a journey. For example, we might need to take a taxi to the airport, plane to another city, and train from the airport to downtown. A travel service should recognise and anticipate our next needs, so when we land at an airport, the system should check whether the next travel has been arranged, and propose means to get from the airport on to the next destination. Of course, the system should be intelligent enough to understand that if we are continuing on another flight right out of that city, then there is no need for travel to downtown.

6.2.2 Mobile advertising with scheduling

Connecting travel offers some opportunities for alternate local travel means to promote themselves. In some city a speed bus service might want to advertise at the connecting page to offer their service rather than the taxis and subway that might also serve the airport. In many cases, the traveller does not know multiple means to get to the hotel, and might take the first one available.

Operators can provide entertainment for the time that it will take for the bus to arrive, and even an enterprising taxi service might offer promotional ads to all who ask for a bus arrival whenever the service says the next bus is expected to arrive later than 10 min from now. It may be that such a commuter would normally not consider a taxi, but if one were offered, would actually take the taxi.

Hint for mobile advertisers: look for competitor's problems.
The 3G environment will have an abundance of services; your advertiser may find best business near to its competing offering, such as the taxi and bus example.

6.2.3 Branded tourist guide

The modern mechanisms of delivering tourist guide information have serious limitations. The printed books by Fodor's and Lonely Planet Guide and other such publications are excellent sources of information on a given city and help the tourist find hotels, restaurants, museums, sites to see, and shopping, etc. However, as much of the information is changing constantly, for example new restaurants open and old ones close, the printed guides start to become obsolete within weeks of their publication date. There is a big business in reprinting and updating these books, and whole sections of shelves at most bookstores are devoted just for tourist guides.

The fixed Internet versions of the guides, as well as many tourist web pages maintained by city tourist bureaux, etc., are an obvious answer to the out-of-date problem, but a tourist usually cannot carry the fixed Internet as conveniently in the pocket, as one can carry a paperback tourist guidebook. Moreover, not all content of all tourist guides is available for free on the fixed Internet.

The next step is of course porting the fixed Internet and printed book versions of the tourist guides to the mobile Internet. The mobile service tourist guide can actually deliver much more valuable benefits to the user than can books and the current free fixed Internet tourist websites. The key is of course the 5 M's. Now the tourist guide publisher has a natural way to be paid just like they get paid when publishing books, and the mobile Internet versions of their guides can generate extra revenues which the fixed Internet versions failed to deliver.

6.2.4 Location-sensitive tourist guidance

The Movement attribute will be the driving element in providing tourist information. Whenever the network notices that the mobile phone has travelled beyond the home-town, then tourist guide information would be made available with only a few clicks, and of course always offering the current location information at the top of the menu options for tourist information. Therefore, when the network detects that I have landed in Rome, it would create one link on the top of the portal window, with the link to Tourist Guide Rome. I might be very familiar with Rome and not need the info so the page should not be pushed at me, only a link. Equally, I might have arrived in Rome for the first time ever and perhaps had not even thought about buying a tourist guide. The ability to access Rome-specific tourist info right then when in Rome, would be particularly useful to me. It is even possible that I did not know I would be coming to Rome when I left home, such as a change in business travel plans, airline delays, spontaneous changes, etc. In that type of instance, if I happen to arrive in Rome very late at night, it might be that the *only* information I have upon arriving in Rome could be the mobile tourist guide.

6.2.5 Time-sensitive tourist guidance

The Moment attribute can be used to advise on time-of-day and calendar time related options, for example proposing theatre tickets during the afternoon, but not anymore at midnight when the theatre plays will have closed for that day's performances. However, at midnight the service could still offer listings of late night restaurants and nightclubs, etc.Then in the morning, it could suggest jogging routes, early morning coffee shops, etc.

6.2.6 Personalised tourist guidance

The Me attribute would be particularly strong in tourist guide value and customisation. Of course, in most cases, the guide could be built to show local hotels and restaurants and banking cash machines, etc., of universal interest to most tourists. Nevertheless, the service for me should also cater to my known and stated interest.

So if I profess to liking live football, claim to be an avid Rolling Stones fan, collect stamps and prefer Mexican and Chinese food, then the tourist guide should automatically show me if there happen to be live football games in the town, and if tickets are still available; keep track of Rolling Stones-related curios, such as that if the local Hard Rock Café happens to have a guitar or other gear of any of the Stones. If yes, the Hard Rock Café and its Rolling Stones affiliation would be mentioned in my guide, but if the local Hard Rock Café did not have Rolling Stones memorabilia, then there would be no mention of the whole location in my personalised tourist guide. The guide would plot out the stamp sales and collecting stores in town. In addition, of course always edit the local restaurant listings so that it would first show only the Mexican and Chinese restaurants close by, and only after those, offer other restaurants.

Hint to 3G network operator: do tourist services early.
Travelling businessmen have urgent needs, often are first to have the terminals and are not as sensitive to price as your local customers.

In this way my personalised mobile Internet tourist guide could be totally different from that of a close colleague whose current interests could be around her young daughters and their Barbie phase; whose music interest would cover the Corrs and the Cardigans but not in any fanatical way as to go visit a place with their memorabilia; whose primary passion would be Ally McBeal and this again would only govern the one day TV schedule; and whose dining preferences would be Italian.

So, my colleague, upon arriving in Rome, would see in her tourist guide totally different information than I would. She would see the toy stores with Barbie dolls and accessories, an indication if Ally McBeal happened to play that day on local TV, and her restaurants would suggest Italian food. Yet both of us would feel we got much more out of the mobile Internet version of the tourist guide, than anything that might be possible in a printed guide or through the fixed Internet.

6.2.7 Tourist guide service evolution

A 3G service will evolve with time. The tourist services could be offered today on SMS (Short Message Service), WAP (Wireless Application Protocol) and GPRS (Global Packet Radio System), and evolve towards the 3G service offering. The service could grow adding sophistication and utility.

Growth of tourist guide
Location-aware pull information
Location-aware pull information with location-sensitive map
Location-aware pull information with map and guidance
Preference-based pull information with map and guidance
Preference-based push information with map and guidance
Time-sensitive push information with map and guidance
Preferences-learning intelligent guide with map and guidance

In the above example the service is constantly a location-aware tourist guide, it only keeps getting more user friendly at every step of its evolution. The network operator should not wait to make the service 'perfect' but rather launch it soon, and learn from customer feedback what to do next to make it better. Adapting to needs will be more important than attempting to 'guess right' in the first place.

> **Hint to 3G network operator: make sure your service creation platform allows fast updates.**
> The ability to modify services to react to market pressures will be a key to success in the marketplace. Make sure your service creation platform allows easy modifications, updates, and revisions to your services.

6.3 Translation services

Translation services are another huge early opportunity. The science fiction example is Star Trek's Intergalactic Universal Translator, which instantly converts any galactic language to any other. That idea is probably several decades away from today, but simpler versions of translation services are being developed, and some simple versions already exist today.

6.3.1 Typed text translators

The first and easiest example is typing and translating. Such automatic phrase dictionaries and intelligent translators already exist covering a wide array of languages, and some of the most common languages can even be translated for free on the Internet, with written languages of English, German, French, Spanish, etc. being already covered.

A simple tourist guiding translation service could be that the user accesses a translation service with the mobile phone, types in the phrase in his native language, and the network translates it to the local language. Then this can be shown to the local merchant, taxi driver, etc., to communicate. The current systems have problems with syntax, at times select the wrong meaning out of words with multiple meanings, etc., but as processing power continues to grow, and the logic built into the translation systems keeps getting better, and the libraries of syntax keep getting better, these kinds of early translation problems will diminish and disappear.

More sophisticated services could be built to cover instances of, for example, the Chinese ideogram character set, which does not conform to the same alphabet set as used to write words in the Western world. Some such English-to-Chinese 'drawing' translating programmes exist also at least for the PC.

6.3.2 Translate text and say the words

The evolution of these would include text-to-speech conversion that would allow me to write in English, and the system to translate the writing into French, and then pronounce the same phrase in French. We are not at the intergalactic universal translator, but taking small steps towards it. Moreover, for most tourist uses this would be a great improvement over current miscommunication and non-communication in many strange countries.

6.3.3 Human interpreters

An interim solution is to have human interpreters available that can be called. They listen to what you say, and then translate it and speak it to the phone. You hand the phone to the taxi driver or store clerk and thus translation takes place. Services like these exist in many countries in Eastern Europe, Korea, etc.

6.3.4 Translate my speech

The preferable evolution will be the combination of speech recognition, translation and text-to-voice. There are large numbers of software companies working on getting the speech recognition systems to function on a PC type environment and to reasonably small errors. In the not so distant future speech recognition will allow us to speak to a device and it to understand well enough to transcribe everything we say. Linking that to translation and text-to-speech, brings us a written translation service, where I speak to my 3G phone, it accesses a network-based translation computer and language bank and syntax library, and in a very short moment speaks out the same sentence translated to

the local language of where I might be. The great benefit is that the caller
deals with an automated device in the network, which allows for profitable
services from automation. This is not a far-fetched future, first such services
have already been introduced in Japan on the KDDI network, translating
across Japanese–Korean–English.

The processing power and storage ability of mobile handsets will make it
quite unlikely that we would carry fully fledged translation services built
into our mobile phone. The required processing power and the large databases make translation services naturals to run off network-based servers.
The network operator would have a major role in enabling and developing
the systems, but will need top expert translation specialist companies to do
the 'content' for the service(s). The 3G network operator can provide valuable information to the utility of the solution, such as the personal information of the user and preferred language(s), and the location information of
where the person is, for the local language options. Still the majority of the
money is likely to go to the translation service.

6.4 Business services around Movement

This section focuses on that type of business applications where Movement
attributes offer the greatest benefits. In some way, almost all of the services in
this book are business services if considered in the context of business-to-
consumer (B2C) but there are many which have significant benefits within the
business or in business-to-business (B2B) and business-to-employee (B2E)
activities. B2E will be covered in more detail later in this chapter.

6.4.1 Fleet control

Perhaps the most obvious Movement attribute business efficiency solution is
fleet control. This could be trucks, taxis, ships, bicycle couriers, motorcycles,
etc.; any kind of fleets of vehicles that move. For trucking, parcel and package
and moving companies the common problem is that the needed truck is
always in the wrong place, or moving empty from one location to another
while a package is waiting to be delivered on the same route.

Fleet control solutions will allow collecting real-time data on where the
various elements of the fleet are, their status and load, driver availability, etc.
For example, there might be a truck in the right place, but it has just been
found to have engine problems. Or its driver might be in need of rest, and
another driver needs to be brought into the location. The benefits to a ship-

ping, trucking and parcel delivery industry are obvious, and similar benefits can be achieved in any wholesaling and transport-heavy industry. A wide range of computerised distribution management systems exists, and these are being ported now on to mobile-enabled and mobile-controlled versions. Many shipping companies already use various wireless tracking and reporting systems. The best-known early mobile adaptation of such a system is DHL's WAP-based package monitoring solution.

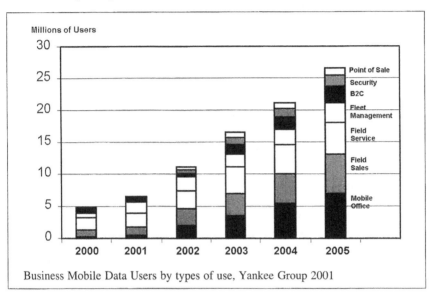

Business Mobile Data Users by types of use, Yankee Group 2001

The money aspect of such services depends very much on the degree of the integration work that is done. If the 3G network operator does part of the integration work and combines fleet scheduling information with other information and communication systems of the company, the 3G network operator could take some money out of the fleet management solution. But if the system is built on the data traffic ability of the 3G network, and the integration work is performed by IT specialists, then the 3G operator's role would be limited to data transmission and location information.

6.4.2 CRM and the corporate portal

Next consider an example of a business sales representative at his customer's premises. Ideally, productive sales reps would prefer perfect information systems from the office, and yet be able to be at the customer site constantly.

Businesses have been building CRM (Customer Relationship Management) systems and been integrating various billing, sales, contacts, etc. databases into the CRM. CRM will be discussed in more detail in the Me chapter (Chapter 8), but CRM's use with the corporate portal will be discussed here.

3G services allow for more integration of the corporate systems to bring the benefits directly to the travelling sales representative. On his phone he will have the ability to connect with the calendar, have schedules made and revised on the go, and read e-mails and make contacts while travelling from one customer site to another. A big key will be giving access to the CRM from the personalised portal set up by his employer's portal managers. Through the portal, the busy sales rep can tailor his phone interface to include only those information sources, which he needs, and exclude all that would create clutter and information overload.

Hint to 3G network operator: get strong IT integration partner.
These types of multi-platform systems are very complex to integrate, get a strong partner early and make sure they will remain with you for the long haul.

6.4.3 Order entry system

Prior to the visit to the customer, the salesman may consult the customer database to obtain background material concerning the customer. The purchasing history may be referenced, and compared with forecast information. The online sales and marketing application may use this information together with current stock levels and production information in order to formulate real time and personal promotional offers. The information available to the system would ensure that customer needs were met with maximum profitability.

Such a combined CRM and portal system would bring numerous benefits. The marketing messages and campaigns that are targeted to that customer can be provided for the sales representative, so that he is not surprised if the customer brings up some ongoing campaign. Orders of course should be made as easy as possible to process, ideally so, that they could be processed online at the customer's site. This eliminates the delays and errors of post-processing at the office at some time after the customer visit. The sales rep should have instant access to production schedules and stock levels to confirm delivery dates. From the customer order fulfilment point of view, the earlier capture of the order would allow for shorter delivery times and better customer satisfaction or competitive advantage.

Of course with this direction of integrating data collection and process control systems deeper and deeper within a company's core functions, the CRM integration work approaches that of process re-engineering. CRM and process re-engineering are themselves a whole science of management consulting, and numerous volumes have been written about these. Probably a point in time arrives when the 3G operator's ability to provide deep process re-engineering ability will end, and management consultants, efficiency experts and IT specialists are needed to fine-tune the system.

The parties making the money in these types of solutions are mostly the system integrators who connect the databases, the access devices and the mobile service. This can be the 3G operator, but could also be a service provider, an IT consultancy, or a specialist systems integrator. For the service revenues, the 3G operator would probably mostly only get data transmission revenues from this type of service, and the company that built the integrated system would take more of the total cost.

6.5 Telehealth services

Another major area of services, which benefits particularly from the Movement attribute, is that of healthcare services that utilise telecoms solutions, so-called telehealth services. This is a very specialised technical area of medical instrument and procedure development, and obviously, a scientific discussion of telehealth would require the treatise by experts in this field. For the purposes of this book, some major areas can be mentioned.

6.5.1 Patient monitoring devices

Telehealth services include remote monitoring of patients. Actually some very basic such systems have already been deployed on fixed and mobile networks for many years. The 3G network and the intelligence that can be programmed into remote monitoring devices opens up a great potential for monitoring systems that can be portable, allowing patients more freedom to move while still being under care and supervision. The added convenience of video and data connectivity allows transmission of status reports, measurements, and even live pictures to show what is going on.

6.5.2 Remote care

Care in emergencies is another area, which provides a lot of potential from the 3G network. Expert doctors could be reached almost anywhere in the world, and images, X-rays and test results could be sent via the network to a 3G device that the expert doctor would carry. Then of course the remote expert doctor could use the 'Show Me' feature to ask the local administering doctor or nurse to show close-ups of the patient, etc. For more on Show Me services, see the Me chapter later in this book (Chapter 8).

6.5.3 Access to doctors

While we often feel we want to physically meet the doctor and actually show the doctor what is the matter, in some cases we may prefer to have remote video access to the doctor rather than not getting to meet at all. For example, this could be if we have a doctor who is familiar to us, and we happen to be out of town. Or alternately, it could be that our doctor happens to be on a seminar trip out of town. Remote access to doctors would be a video call to the doctor, which could easily include the cross-transfer of X-ray pictures, test results, etc., which the doctor could show via the 3G device, and discuss at the same time on a live voice connection. As the regular 3G phones would be small and limited in resolution, screen size and so forth, the hospitals could invest in special remote access devices with larger screens, etc. to facilitate this type of contact between patients and doctors who are situated far away.

Telehealth is a special area where manufacturers of medical instruments will be bringing a lot of their expertise to the partnership with the 3G network operator. Much of the money to be made is in the manufacturing of the custom devices and in the integration of the systems. Special care will need to be taken to design 3G terminals and hospital devices to function with a minimal of interference so that they do not cause disruption with other electronic diagnosis devices, which hospitals already use.

6.6 Synchronising gadgets

A major area we will find to be a nuisance, and one which many device manufacturers will try to solve, is the harmonisation and synchronisation of our gadgets. As an example, let's assume I make a purchase on the fixed Internet from my PC at the office. The overall network should be intelligent enough to recognise that it is still I if I contact the same store from my 3G phone and add to the order, without having to re-enter me as a new user when I happen to access the service from a different device. Then if I use my 3G system in my car to contact the same company to change the shipping address, again it should recognise me and not make me identify myself a third time to prove that I am the same person. The 3G network has the ability to associate specific devices to me, and control my personal data centrally, so solutions can be built to make a certain 'family' of gadgets to 'belong' to me and be seen as controlled by me.

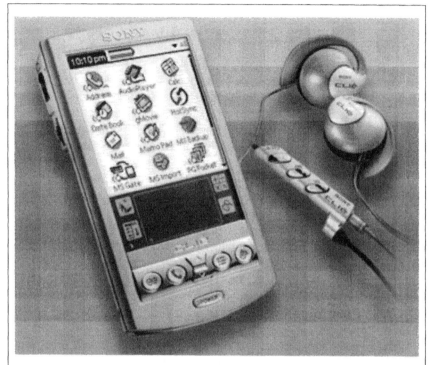

Personal Digital Assistants (PDA) are becoming more popular and are integrating browsing and telephony with traditional personal organisational features. The Sony Clié is one such example.

6.6.1 Virtual (network-based) PDA

Another area, which will need to be designed to recognise the Movement attribute, is the various PDA (Personal Digital Assistant) functionalities such as calendaring, scheduling, to-do lists, phone books, etc. As the technology and device manufacturers come up with ever more gadgets to help us in our daily lives, the synchronisation of those devices becomes an ever-increasing problem. We all have experienced it at some level, where the calendar entry is in the calendar in the PC, but not in the PDA, or the person's phone number is in the office phonebook, but not in our personal mobile phone phonebook, and again the person's e-mail address is on a printed business card but not in the electronic system. Another example is when we cross off an item in the to-do list on our mobile phone; it does not disappear from the to-do list on our PDA nor the computer.

There will be many solutions developed to cope with these issues. However, at the very heart of the problem, this type of synchronisation is not so much a *computation* problem, but rather a *communication* problem. The devices need to communicate with each other to update changes. One intuitive way to resolve this is to set up a central location where to store the changed data, and have all devices access that central location. In this model the natural owner is the 3G network operator, which would house individual virtual PDAs for their users, and allow just about any conceivable device that the user controls to access that virtual PDA.

To use an example, if I use my mobile phone to receive a call, and notice that my friend has gotten a new phone number, I can store that number. While I would store it on to my mobile phone, the system would also store it for me on my virtual PDA. Then when I would access my contact list from my personal computer, that would again access the same virtual PDA file, and I would see my friend's new phone number there, without having to re-enter it. As my car would have a built-in 3G phone, it could also automatically then update its phone list from the master virtual PDA and if my friend calls me later in my car, his name should now display also on the car.

Today many companies are working towards solving the synchronisation problems, and we will soon have solutions. Initially there will be many competing solutions and probably for many years to come several will be popular for their own set of features and functionality. A typical alternative solution is the master gadget solution, where one gadget is the holder of the central data, and others are synchronised to it. This type of solution probably appeals to the 'propeller heads' of the early adopter IT-savvy techno-geeks, but could be quite too cumbersome for mass-market adoption. In synchronisation solutions, the synchronising agent will be the primary one making the money. As more and more of the devices to be synchronised are 3G devices, so too will more and more synchronising traffic be generated into the 3G network, bringing that traffic and revenue to the 3G network operator.

6.7 Services for employees

The trends to move employees away from the office through remote work, etc. arrangements, provides ample opportunities for 3G network operators to provide 3G services for employers. The employers have a new need to manage their employees who may work from unconventional locations at odd hours. These kinds of services fall under the B2E category. B2E services include solutions for such trends as having employees working remotely from home, closer to the customer or even at the customer, on the road such as at airports, hotels, etc., and part-time and flexitime work, and even working for multiple employers. The different types of mobile users were illustrated in Boeing's survey of its mobile workers.

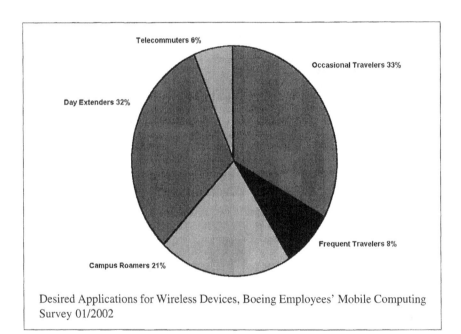

Desired Applications for Wireless Devices, Boeing Employees' Mobile Computing Survey 01/2002

6.7.1 Monitoring employees

One of the major issues affecting an employer's ability to manage remote employees, and for providing various costly services to them, is knowing when they are working. In some countries, this is having taxation effects, can your work mobile phone be used for personal use, and if so, it could be taxed as a benefit, etc. The 3G technology allows new ways to determine whether an employee is working, and/or allow the ability to access certain of the employer's resources only when working, and from designated locations.

The location information is the first and most obvious attribute. The company's sites, such as offices, warehouses, factories, etc., can all be designated as 'work' and thus anytime the employee is at any of the company's locations, the employee would be considered to be working. The employee's home could be designated as a partial workplace, where for example office hours could be enforced. So, if I happen to be at home on Tuesday at 10:00 in the morning and place a call, I would be considered to be working. But in the evening, if I placed a call from home I would be considered to be not working. The third way 3G networks could monitor working would be by using the status attribute. I could be allowed to tell the corporate system that now I choose to work – for example on a given Saturday from home – and in that case, again I would be recognised as working.

The benefits of determining if one is working or not are considerable. First, they can be used to track the hours for workers who need to report hours. For example if I am supposed to work from the office or from home, and I happen to be visiting my friend, then the system would not accept that location as being at the designated work area. Secondly, the system would allow for controlled access to corporate resources. For example if I am allowed to access corporate resources like the corporate intranet only from the office or from my home, then if I lose my phone, the person finding the phone cannot access the corporate network because the phone is in the wrong location. A third benefit would be for accounting and tax purposes, so that any time the phone was used in the evenings or weekends, not working, would not be work, and depending on how such benefits would be provided by the corporation, the employee might have those calls deducted from the paycheque, etc.

6.7.2 Secure access from the road

Another major concern is the secure access to corporate intranets and other corporate data, while the employee is travelling. There are numerous existing solutions to handle the passwords and data encryption, etc. The 3G network operator will be a natural contact point for developing the current cumbersome systems into more simplified yet more secure solutions. Most of the basic office efficiency tools that we use on a PC, such as e-mail, word processing, PowerPoint, and spreadsheets, can be run on handheld devices such as the Blackberry, the Nokia Communicator, various PDAs, etc. There will be top-end 3G phones which will incorporate these kinds of benefits, where in some cases employees might abandon the notebook computer altogether for a handheld device, and at least in many cases the primary office tool for travelling workers would be the handheld 3G device rather than the notebook computer. The most desired service business people want on the mobile phone is access to e-mail. In fact the first study of business users given a choice of using a handheld device was conducted by Goldman Sachs in the US. It found numerous benefits from mobility, and that users preferred their handheld devices, and that 20% of the users stopped using their laptops altogether. While this is the first reported study of such trends, probably numerous studies reporting in 2002 will report similar findings.

When the 3G corporate solution is built to include the 3G handheld devices and provide secure access to those, the 3G service becomes the natural platform for all remote access and secure access related needs of businesses. The 3G network operator wishing to build this ability will need strong IP (Internet Protocol) and telecoms security know-how and might want to engage a partner or partners in this area.

6.7.3 Bundling home and office services

The 3G network provider could also provide a combination of business and residential services via the same phone. So by using the status of am I working or not, I could be given access to my employer's services, or during my free time, I would get a special benefits package offered to all of the employees by the 3G network operator. So for example, if my employer has a bulk discount scheme for cheaper international calls, there is no reason why the 3G network operator could not offer the same discounted rate to me as a special benefit to the employees. It would need to be agreed with the corporate customer of course, and there might be national restrictions or tax implications from such bundling. Nevertheless, it could be a considerable benefit in a portfolio of services offered by the operator to its corporate customers. Those types of services would then be used by that corporate customer in its benefits package communicated to its employees, amongst the salary, bonuses, healthcare, etc. benefits.

The status of the employee would also serve to steer incoming calls and contacts. So for example if I am at the status of 'at work' – regardless of whether I happen to be at the office or at home or at the airport, or even playing a round of golf, the service would treat me as working. Incoming work calls would be processed accordingly, e.g. the phone would ring. Then when I was not at work, the incoming calls and contacts would be treated in a different way, such as going to my voice mail or to my secretary, or routed to my colleague, etc.

The benefits to the 3G network operator are obvious, more traffic and another way of binding the corporation ever more tightly to that 3G network operator. The benefit to the employee would be convenience; the employee would not need to have another phone handset for personal use. With the advent of 3G, the early handsets are likely to be more expensive than current phones, and thus there would be some benefit also from having the new phone with its added features, to use for personal use as well. Depending on the model, the 3G phone could be the family's first digital camera, or it could replace the worn-out music player/Walkman. The employer gains not only from bringing a benefit to its employees, but also in that the employees will be carrying their new 3G phones with them. As the employees would have continuous access to their calendars and schedulers, etc., the inevitable 'working while not at work' will happen, where employees will do some work also in their free time.

The 3G operator needs to make strategic choices, to go deep into business customer systems or not. This kind of involvement in the business customer's internal IT systems requires also a long-term commitment to partnership with the corporate customers, and the 3G operator will need to evaluate this opportunity carefully. There will be no partial success in this area. Some operators will put a clear commitment into serving the business customers, and build portals, integrate CRM systems, design B2E solutions, etc., and get the business; those operators who do a half-hearted effort will fail. The competitive impacts will be discussed later in this book in the chapter on Competitiveness (Chapter 14).

6.8 My services travel with me — VHE (Virtual Home Environment)

The Movement need is also being addressed through 3G standardisation with the concept of Virtual Home Environment, or VHE. The need for VHE is illustrated by a simple example. Let us assume that a Japanese business visitor travels to Norway. The Japanese visitor does not speak Norwegian. The Japanese person would prefer to have access to his home services, in his own *language* and *alphabet*. The Norwegian operators could, of course, attempt to replicate typical Norwegian services into dozens of languages and character sets including Japanese, and to try to keep them up to date. That is bound to be a solution that will be full of disappointments and mistakes. It is much better to route the Japanese traveller's requests back to his home service, where all the usual information is in its right form, in the right language, and spelled correctly with his home alphabet.

6 Vignettes from a 3G Future: Is the Club Hopping?

We were thinking of going out with my friends and could not decide which club or bar to select. As the weather was bad we didn't want to stand in line very long, and equally, we didn't want to go to a place which was totally empty. A friend of mine took out his 3G phone and dialled the linecams and floorcams of the various nightspots and we saw immediately which club was the one we wanted to go to tonight. I made a note of the numbers of the cameras and will use this trick also the next time I go out.

Digital video cameras that connect to the Internet are becoming commodities of very low cost. Cameras could show how busy a dance floor is, how many people are at the bar, how long the line is outside, etc. Such views from cameras can be used as the new marketing means replacing posters and billboards for various events and entertainment opportunities. Such video cameras could be accessed for free on the fixed Internet, and at a cost from the mobile devices such as 3G phones. The mobile access could be made free through sponsorship for example by breweries.

That is what the VHE is intended to provide. Services on VHE will be inherently Movement services, but VHE does not bring anything 'new' beyond what we have already at the home network and service portfolio. VHE only extends the home services to travel with us. For the mobile operators who are considering the priority of VHE, they are best to keep in mind that the VHE travellers are likely to be early adopters and wealthy – they *have* a 3G phone, and they *bring it with them*. These are very likely high volume callers and users. If you make it easy for them to access their home services, you will be putting a lot of data on the network, and can probably charge a premium for such access as a VHE package. This is very lucrative traffic, and operators should enable them very soon in the introduction of 3G network services.

6.9 Real services today on Movement

The ideas presented in this chapter are not wild fantasy visions of a far-away and improbable future. Many of the ideas exist in limited form today on SMS and WAP-based services on GSM (Global System for Mobile Communications)/GPRS networks, on I-Mode and other Japanese networks, on fixed Internet and broadband services, etc. They will still need to evolve a lot to fulfil the promises and potential discussed in this chapter. Still, clear signs exist that mobile operators today are already making money on many of the same service ideas around the Movement attribute, and consumers are willing to consume those services in their current rudimentary formats.

6.9.1 Underground timetables in London subway

The London Underground or 'the Tube' is a complex system of interconnected routes, and with the service over 100 years old, there are lots of technical challenges keeping the trains modern and up to the transit needs of one of the world's largest cities. There are frequent subway train delays and various information boards indicating the next expected arrival of trains, etc. This information is being provided on a WAP-based service to London commuters and tourists.

6.9.2 Next bus

Similar services are appearing in many countries combining the location of the person and the bus timetable. One of the early such systems was introduced in the US.

6.9.3 Text translation services

Some translation servers exist on the fixed Internet where you can type sentences in German and the system automatically translates it to English – or at least something approximating English. These are still developing, as there are still numerous problems with vocabulary, grammar and usage.

6.9.4 Spoken translation services

Voice-oriented human translation services exist in many countries where you can call a translation service, speak to it, and there is a human interpreter who translates the service on the phone. In Japan the first machine-based translation service has been introduced where up to 7 s at a time can be spoken and the machine translates between Japanese, Korean and English.

6.9.5 Airline scheduling notices

Several airlines offer SMS text message and WAP-based information to frequent fliers and other passengers who sign up to their services. Swissair was one of the first to offer these types of services. Gradually airlines are expanding the offering of services adding things like accessing frequent flier miles programmes, viewing flight schedules, etc. Recently Sabre announced a system to allow numerous airlines and travel agencies to offer such a service.

6.9.6 Location-based tourist information

In Italy, the mobile operators offer tourist guides that are location-aware. They recognise if you are at a statue or bridge or museum or other tourist attraction, and by sending a request to a number, the network returns information on that tourist item, such as what is it, who made it and when, etc. The network identifies the accuracy of the wandering tourist within about 50 m.

6.9.7 Location-aware maps for tourists

Japan was among the first to introduce location-aware maps. Other variations include giving written or spoken walking or driving instructions via the mobile phone.

6.9.8 Location-based advertisements

Several shopping centres in the UK and US have set up systems where the shopper can register to get ads when the shopper is at the shopping centre. These are all permission-based ads where the user signs up for the service.

6.9.9 Wholesale catalogue/pricing

The British car prices and parts company, Autoglass, has brought its service on to WAP, and rather than scanning through thick printed used car parts and costs catalogues, car dealers and repair centres can now access the information directly via WAP phone.

6.9.10 Remote patient monitoring

In America a company called Life Alert has for several years offered a simple wireless alarm system by which home care patients, such as elderly people, can contact the company for emergency assistance.

6.10 Moving on up

This chapter has looked at the revenues and profits of some services which have a strong benefit on the Movement aspect of the 5 M's of 3G services. This chapter is not an exhaustive listing of such services, and described mainly only some illustrative groupings of services to provide an understanding of the money side of 3G services that are expected to be created. The chapter has discussed who are positioned to be making money early on with these new services, as well as some of their target users. The reader should keep in mind that each of the described services would have several of the other 5 M's as a strong attribute as well.

The following chapters will examine the other attributes of the 5 M's and show services that can benefit on each of those dimensions. Remember that also with them, most have a significant Movement dimension, and many of them could just as well have been described in this chapter. But most of all, services should be launched and trialled, then improved. Nobody should attempt to introduce a perfect 3G service from the beginning. In designing 3G services the operators and content providers will need to experiment and be creative. Perhaps Arthur C. Clarke can guide us with his advice: ''The way to define the limits of the possible is by going beyond them into the impossible.''

7

The Profits of Moment Services:

Expanding the concept of time

The Moment attribute is the most activating of the 5 M's. The ability to expand the concept of time includes aspects such as postponing, doing things with idle time, following up on missed issues, catching up on time, multi-tasking, and so forth, as was described in the chapter on 5 M's previously in this book (Chapter 5). As Moment is the strongest motivator of the five attributes of mobile services, and thus the most powerful catalyst for users paying for services, Moment is of particular interest to content providers and the 3G network operator. If we have a momentary need to get some information, we might be willing to pay a considerable premium on the instant access to that which we desire.

The services described in this chapter have a high benefit on the Moment attribute, but as in the Movement chapter (Chapter 6) and the following three chapters, any 3G service should be built to utilise all of the 5 M's. There are hundreds of services that are strong on the Moment attribute and this chapter has only a chance to explore a few of them.

Moment is the most powerful activating catalyst of the 5 M's. When one has a sudden urge to gain access to a service or some information, it can be a very high value need and in extreme cases the Moment attribute may have almost limitless potential to activate. To give an illustrative example, if you are on your way to an important business meeting and accidentally have your clothes soiled, for example by a passing car running through a puddle and splashing dirty water on you, then you have a sudden need to know if there is a clothing rental place, or a clothing store nearby. While normally you would not pay anything to find the **location** of the **clothing rental store**, or a clothing store, in this Momentary need you might be willing to pay very much to satisfy that temporary but urgent local information need. If someone could catch you at that very precise moment, the person could perhaps sell you information of the clothing rental location a few blocks away, for several dollars, perhaps up to 10% of the value of the new suit if your meeting was important enough.

It is important to note that normally we would not pay anything for knowing where the nearest suit rental shop is or other such information. But this example illustrates the temporary way how the Moment attribute may exceed all 'reason' and produce opportunities to sell valuable services.

Obviously the 3G mobile operator cannot build its business case purely on occasional clothing rentals, nor is it even likely that the 3G service would be able to capture all such sudden needs. In the near future, however, an ever-increasing number of these Momentary needs will be met by mobile services. Most importantly this sudden need can be a very powerful catalyst for action – and motivator for payment. The prudent 3G mobile operator will build services which serve the needs to manipulate time, to stretch time, to re-schedule, to make use of idle time, to multitask, etc.

The Momentary profit opportunities are greatest when the emotional factors are high and thus 'reason' is low, for example attending a wedding or attending the championship game, etc. The utilisation of Momentary services depends on awareness of the service's existence and the trust in it functioning properly. The potential is great for building compelling services, but it will take a lot of time, years, to create the change in behaviour where the service becomes natural for consumption via the mobile phone.

7.1 Mobile entertainment

A Moment need can easily arise whenever we have some time to kill. This could happen at the bus stop, the airport, standing in line, waiting for a friend, sitting in a taxicab, etc. These spare moments are opportunities to relax a bit and laugh a little with a joke or cartoon, or play a game, or catch up on the sports scores, or whatever brings entertainment or relaxation to a spare moment. This part of this chapter will look at several general entertainment areas, while adult entertainment will be discussed in the Me chapter (Chapter 8).

The entertainment industry is becoming digitalised and has numerous aspects which transform well to the new 3G environment. The music industry is looking forward to the chances to have music streaming and downloading directly to 3G terminals which would also be music players such as MP3 players. The gambling and betting industries are one of the most profitable parts of the fixed Internet, and are looking to repeat their success when the gambler and betting person can have instant and continuous access to any sporting event and betting situation. National and cross-national lotteries are another area of similar interest. There are hundreds of industries related to entertainment which are all looking at entering the 3G environment, from small jokes sites on the WWW (World Wide Web) to the global entertainment empires such as Disney. The users of mobile entertainment are expected to grow at a very strong pace over the next few years as the ARC Group's study shows.

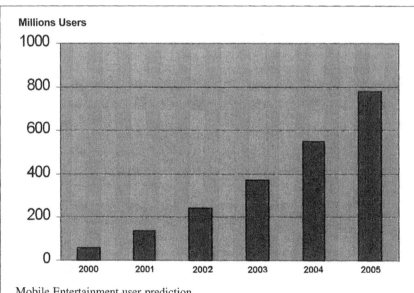

Mobile Entertainment user prediction
Source: ARC Group Mobile Entertainment Report, June 2001

7.1.1 Micro movies and advertainment

One of the big promises of 3G is video on the phone, which is understood by many to include video calling, television on the mobile phone, and movies on the mobile phone. Actually the bandwidth and data transfer rates of 3G do not readily support the viewing of television or movies on the phone, even streaming music is going to be relatively costly on 3G technology. Video calls will require reserving a lot of capacity per call, using a high QoS (Quality of Service) class. But the mobile phone will be getting its own new video entertainment. That will come in a new form called 'Micro Movies'.

Micro movies are special small-screen format short-duration films being developed by the motion picture industry in Hollywood. Several early micro movies already exist. The films are planned to be roughly 5 min in length and thus easy to consume when we have a moment of extra time. These are not 'trailers' that currently Hollywood makes to advertise normal 'long' motion pictures, but rather complete stories told in 5 min, produced for the small-screen format.

For those who might doubt the viability of a 5 min story, one must remember that Music TV (MTV) **music videos** often tell stories, and typically run 4 min in length. A short story can easily be filmed for 5 min duration. The micro movies need to be directed with more close-up scenes of the actors showing their faces much more than in typical Hollywood movies, and of course epic-like scenes of thousands of warriors or aeroplanes fighting in the sky, etc. which work well on the big screen, and also will work on television, will not work on the small-screen format.

For early adoption of this service, the best aspect of the micro movie is its pricing structure. The concept of the micro movie is that they will be free for the viewer, totally paid for by advertising that goes with the 5 min film clip. The advertising is likely to be embedded into the show, so that there will not be 'ad breaks' in the film clip, but rather the items and their advertising will be similar to advertising/sponsorship on Hollywood movies. Some have started to call the merger of advertising and entertainment by the term 'advertainment'. The revenue generation to the 3G operator will not be much more than merely delivering high data content, but the perceived value to the user will be great. The analogy for the service is much closer to commercial TV than Hollywood movies. This means that the micro movie format will likely work well for serial content, such as daily **monologues** of **TV comedians** such as David Letterman, Jay Leno and Conan O'Brian, short daily **updates** of **soap operas**, news updates, sports roundups, etc.

For the 3G mobile phone user, the advent of micro movies will give them a lot of value in new and exciting entertainment that they can consume on their phones, without bringing additional costs to the user.

7.1.2 Entertainment bundle example

Successful 3G services will rely on techniques of bundling and segmentation. Typically a bundle offered for a given segment might have a set of standard components and optional components for a set, relatively small fee. In this example it is assumed that the service bundle would be a basic simple entertainment package, with **jokes**, **horoscopes**, puzzles, games and cartoons. Typically the content would exist elsewhere already in digital format, and the owner of the content would join in the bundle. Various syndicates exist to provide aggregation services for the media – especially newspapers around the world, and similar concepts probably will emerge for 3G.

The users would get to select which services they want to have immediate access to on their portal on the small colour screen of the 3G phone. Those services which are most attractive to the particular user would be the ones that the user would select. Typically this could be only the most favourite few cartoon strips daily, and for example only the user's daily horoscope rather than seeing all 12. The service should allow the users easy access to the other entertainment content, but not force the users to do any unnecessary scrolling or hunting for favourite pages.

Each of the content providers in the bundle would have some kind of revenue sharing system with the bundle, partially depending on the popularity of the service, and partially depending on the network costs of delivering the content to the user. For a small text-based joke, or even the simple graphics of a black-and-white cartoon strip, the actual load to the 3G network would be very small.

As one example, let us assume that the basic entertainment bundle would cost 1 dollar per month, and that it would appeal to half of the subscribers. In a typical large European country in 3G the operator might have 10 million subscribers, and if we assume half of the users would take the basic entertainment bundle, this brings 5 million users of the basic entertainment bundle. The bundle would generate monthly revenues of 5 million dollars, or 60 million dollars annually for the 3G operator. These revenues would need to be split with the content providers. Let us examine how that revenue might be split in the case of one of the content providers.

Note that the operator would have hundreds of such content providers in this system and each would get part of the revenue sharing. The overall billing of this service would generate 5 million dollars per month, or 60 million dollars per year, with minimal operator involvement beyond setting up the links to its billing system which is built to track billing worth pennies anyway with the voice minutes traffic. Most of the revenues would typically be returned to the owners/creators of the content, which is the value to the user. Reasonably the operator could be expected to retain depending on service anywhere from 10 to 50% per service, depending of course most of all on the loads generated on the network, and the depth of information needed per user from the billing system. We can assume for the sake of argument that the 3G operator could keep on average 30% of the revenues through its value-add in charging and billing, its work in aggregating the entertainment bundle, and from including the content providers into its standard entertainment bundle. Thirty per cent of the revenues from this entertainment bundle would bring the operator 18 million dollars per year in this country.

By giving the operator 30% of the revenues of the entertainment bundle, it would leave 70% of the revenues to be split with the potential entertainment bundle contributors. If each would get 1/10th of a cent per subscriber to the bundle, it would allow a maximum universe of 700 potential content providers that one could select from. This could include for example 70 separate

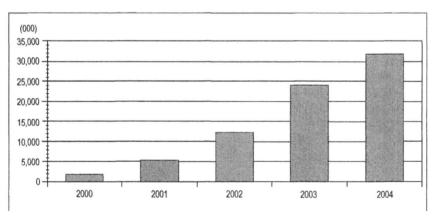

Source: IDATE 2001.
Music will be one of the major categories in the entertainment segment and will gradually capture a growing percentage of the total music sales market. Downloading music to new entertainment type UMTS devices will expand as the cost of data memory prices fall.

cartoon strips, another 70 separate daily joke services, etc. Early on, from a practical content point of view, this might be a selection of 100 or 200 services, gradually expanding to for example 500 selections. The user would be allowed to select any 12 favourite entertainment feeds, which could then be four cartoons, a daily joke, the daily horoscope, two daily trivia games, a brain teaser, etc. etc. etc. Your selection of 12 would be quite different from mine.

In this example each of the content providers would on average get 1/10th of a cent per subscriber. With 5 million users, the revenue would be 5000 dollars per month or 60 000 dollars per year for the content. That would be an average number, the more popular services would get more, the less popular services less. Early on when the entertainment bundle is being set up, the operator could offer the early content providers a guarantee of 1/10th of a cent per subscriber for the first 6 months, and after that start to implement payment by popularity within the bundle.

7.2 Repackaging serial content

The most popular **cartoons** like Dilbert, Peanuts or Garfield are currently available in daily newspapers where they are part of a newspaper and also available for free on their fixed Internet websites. These websites typically show the daily cartoon, but usually with a lag from the newspaper versions, so that today's strip might appear 2 weeks later on the free Internet version. The web pages typically also store a small collection of the previous cartoons, such as all cartoons for the last week or all cartoons for the last 30 days. The websites tend to be sponsored, so the cartoonist is likely to get some advertising revenue from the fixed Internet site. But the cartoonist does not make money on visitors just seeing the cartoon. Of course the cartoon authors get money from the newspaper version of their cartoon, and also sell books with old strips and make more money that way.

3G provides an opportunity to get more money out of that very same content which is already created, once sold, and already exists in digital format. The content provider and network operator should invent numerous ways to generate more revenues out of the very same content selling the same content many times over. The key to this as so many other payments to content providers, is micro-payments, or payments worth less than a dollar. For more on micro-payments, see Chapter 4.

7.2.1 Daily serial content

7 Vignettes from a 3G Future: Temporary Doggiecam

Our puppy has been sick and is still not very well accustomed to us being away some evenings. So I placed one of the family 3G phones with its camera pointing at the puppy's room and set the phone on remote access mode. From the party we can take brief video calls every so often to 'check in' on the puppy and see that she is ok.

The exact features of 3G phones are limited only by human imagination. With many more manufacturers of 3G compliant devices than currently on any given cellular phone standard, the competition and market opportunity will allow for numerous special features and possibilities. Whether a phone manufacturer would want to allow remote activation of its camera would be up to the manufacturers, but the always-on aspect of 3G, and the digital aspects of the design of the terminals will make this type of feature very possible.

Daily serial entertainment content includes jokes, horoscopes, **crossword puzzles**, **trivia quizzes**, cartoons, etc. Most of these are not particularly tied to any given day, with horoscopes being perhaps the most obvious exception. So the crossword puzzle from 2 weeks ago is just as valid as the one in today's newspaper, provided that the puzzle is new to you, and nobody has filled it in or shown you the correct answers.

The content creators need to deliver serial content on a schedule to their distribution channel, mostly the newspapers. There is no technical or business reason why this content could not be offered also on a fixed Internet-based service with a time delay. The biggest reason why this is not so for all such serial content is probably related to the contractual relationships between the content producers and the various content syndicates which provide the content to the newspapers and other media. If the content already exists on the fixed Internet, the new mobile services' opportunity is a wonderful new way to generate revenue for that very same content. The daily serial content would be made available to the 3G users on a daily basis, for a small fee such as 1 or 2 cents per view. Remember that the actual content for a cartoon is very small and users would easily reject prices of 5 or 10 cents per cartoon, puzzle, joke, etc. But 1 cent or even 2 cents would probably be acceptable.

> **Hint to content provider: keep the price below the pain threshold.** That way you will find a vast number of users paying for your content. Don't get greedy and kill the opportunity with a price set too high.

7.2.2 Content archive

One of the obvious extra uses is the content archive. Here users could scroll back in time to see content they have missed, or content they want to see again. Every view of old content would cost the same as viewing new content. It would be important to keep track of which dates of old content have been viewed, so that the user is not charged for seeing 'if I have seen that cartoon already' on older content. But by allowing users to scroll back months and even years of old content, brings small trickles of extra revenue for content which otherwise would not be generating further revenues.

7.2.3 Content fan club

For serious fans of given content, fan clubs could be set up. These are naturals for cartoons, but could also work for given puzzles (perhaps Who Wants to Be a Millionaire) and other content. Fans would have the ability to chat with other fans, get special deals on accessing the content, etc. The content provider could take advantage of this fan base to test ideas and ask for feedback. The content provider could even provide a newsletter just for the fans. Joining the fan club would be like subscribing to any service in 3G, it would involve a monthly fee for example a dollar, and it would give some benefits not available on other means to access the content.

7.2.4 Selling single content from a series/click to own

Another obvious use is to sell content from the series. If you liked the cartoon, or solved the particularly fun puzzle, you might want to own it as a ready-to-frame print or on a T-shirt, mug, etc. This could include selling any official game scores where the user happened to hit the listing of the best scores. Click-to-own buttons should be used and allow for inexpensive purchases of the content, either in high resolution graphics for printing, or as a printout mailed to the user's address. The cost of the content without the printing and sending would be of the magnitude of 1 dollar.

> **Hint to 3G network operators: click-to-own buttons!** Use them. Offer the idea to all of your content providers, find creative ways to own the content electronically, and also to own it on merchandise, such as printed on a T-shirt or mug, etc.

7.2.5 Gift serial content/click to send

Another way to make more money on the very same content is click to send. Make the serial content easily available as gifts to be sent to friends. All of us have experienced the sensation some times that a given cartoon would be perfect for a given friend or colleague. This should be made very easy, so that serial content could be sent via a single click, and then selecting from the phone's number list the target person's number. Within seconds the friend has received the content. The cost to the sender would be in the magnitude of 20–30 cents or so.

This same model would apply to any of the hundreds of content providers participating in the entertainment bundle, although some might not have as broad abilities to further expand their business – it might be difficult for a horoscope author to re-sell last year's predictions.

All creative content producers gain extra revenues from work which has already been sold and already exists in digital format. The operator gets addictive content on to his network which will generate more traffic and revenues and bind the subscriber to that operator's service bundle. And the consumer would gain multiple ways to enjoy content that the consumer wants. It is a classic win-win-win situation. And clearly a way to make money from the mobile Internet out of content which is currently provided for free on the fixed Internet. This is the business logic of small payments and large populations; it is what I call the magic of micro-payments.

7.2.6 Viral marketing of serial content

It is important to remember that every interested person who would visit serial content, is a potential promoter of it. The service bundle should make it extremely easy for anybody to promote that site by sending greetings and even reward members to get more members into this service bundle. A logical use is to send greeting cards from this site, with a link to the site, as a promotion. And members of the fan club should be enticed to get more fans signed up. A typical reward could be 1 month free for the existing fan for every new paying member who signs up, etc. For more on communities, please see the Me chapter (Chapter 8).

> **Hint to content provider: use community and viral marketing of 3G.**
> There will be a lot of forwarding of 'cool' and good information and entertainment, embrace that and get new users and fans. Understand viral marketing and how communities spread information.

7.3 Mobile information and infotainment

News, financial, sports and traffic information are types of services which can be specified when the subscriber signs on to the service. If you work in the financial industry and have a passion for following tennis, then you might easily subscribe to the daily financial news pages and select the tennis coverage of the sports pages. But with all information content, remember the wisdom inherent in the opening quotation by Johnny Carson, that, people will pay more to be entertained than to be informed. If at all possible, capitalise on that tendency, and *add* entertainment value to your information. When at all possible, try to deliver **infotainment**, not only information. If you add advertising, then make it advertainment – again entertaining as well – and you are likely to be better valued by the person receiving the communication and content.

7.3.1 Information push/news push

Typically most of us have some parts of the daily news flow which are of particular interest, and a lot of others which are not. The overall mobile information package will need a lot of ability to be customised by the user, both in general – such as selecting the financial pages and the tennis updates – as well as modifying the profile on the go as the need emerges. It is quite usual that we might have temporary needs to follow a given area of news. Let us assume that you are about to go on vacation to Thailand. In that case before the trip you might be very interested in receiving news and weather coverage of Thailand. But quite likely some time after the trip you might not want further coverage from that region of the world.

A typical Information Push service would offer choices to select from a portfolio of news coverage. Depending on the overall datacoms traffic load on the network, and amount of higher bandwidth data content such as images and video clips, a part of the service could typically be offered as part of the basic news service package. So for example we could have the option of selecting 10 items from the overall news coverage which would be sent as links to the

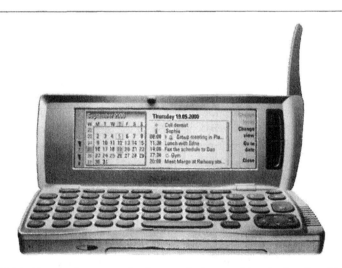

The Nokia Communicator was the first terminal to combine the functionality of a PDA with calendar and contacts directory and the mobility of GSM voice. As the communicator has improved in functionality making it an ideal business tool for many people, traditional PDA vendors are now bringing new product to market to catch by adding mobile voice and Internet capability to their product lines.

mobile phone every day, and we could click on to any of those news items for no additional cost. In this way we would receive our customised – but shortened – electronic newspaper every day. The content would typically be provided in partnership with a major newspaper, TV station, or other news source. As with all cases where the content comes from outside content providers, a revenue sharing mechanism would be needed to cover how the content producers would be paid for their value-add.

7.3.2 Premium news coverage

Of course for those who want more coverage they can subscribe to the full content version of the electronic newspaper, at a cost similar to subscribing to the newspaper in its paper form. And as a third option, on any given day, the daily paper could be accessed at the equivalent cost of that newspaper's daily news-stand price. This would be of particular use for travelling people, who can read their home newspaper at their home through a subscription, but would have the need to read the home-town paper on travelling days, and pay for it also only on those days. For the newspaper publisher, the travelling population is a large missed opportunity on any given day to consume more of the paper. Even those who subscribe to the newspaper are unable to consume it when travelling, and many of these would rather consume the news when fresh, from a mobile terminal, than a few days late when they return home to a pile of old newspapers. The Moment attribute is very acute with any information content which grows old fast, such as newspaper and TV news content. If the news organisation cannot get the content to its readers and viewers very rapidly after publication, the news will grow old and even the loyal customer will no longer be interested in it.

The value of these services is increased if users are able to be kept informed and up to date on subjects of particular interest to them. For instance, users can receive information about the prices of shares and latest news about companies they have intentions of investing in. The service should be created to be very flexible to allow easy addition and deletion of subject matter. For example some investors change components in their portfolio frequently, and may develop sudden interests in given companies or industries. The service should be designed to allow very fast changes on the go.

7.3.3 News partners

With news services the 3G network operator will not be able to provide much of the content by itself. To do reasonable news coverage would require hiring, training and managing a staff of reporters similar to that on a newspaper. This is not the network operator's area of expertise, and it makes sense to partner with news sources such as a major newspaper or TV station, or both. In that case the majority of the value comes of course from the news organisation, and the 3G network operator's role would be modest. However, in addition to delivering the data, the 3G network operator would have benefits from the personalisation of the information as well as the positioning of the news service in the operator's portal.

7.3.4 Operator journalist(s)

The 3G operator may be hiring itself a single local-items editor to cover items such as the local **TV listings**, **local weather forecasts**, the **movie listings** and so forth. This could be part of a very **simple daily news package**, which could also include short summaries of news edited from the major news feeds. When setting up a news partnership with newspapers and TV stations, the 3G operator will need to be upfront with its own news 'organisation' scope and purpose to ensure that this will not become a thorn in the side of the news information push partnership. A good simple guideline is that the in-house journalist would be at home, doing all of the news collection from a desk, collecting and collating from news feeds and such sources. Also such an information offering should be limited to basic subscription minimum offering, and that any 'premium' and value-add services would be provided by the more established news content partners.

7.3.5 Information pull service

Information push services serve needs which we have on a regular basis (such as daily or weekly or whenever events occur in that information area). Pull services on the other hand, are of the type that we do not need on a regular interval, but develop the need suddenly to find the information. It is the difference for example between the newspaper ('push') and the phone book ('pull'). Pull services include most reference sources such as the telephone directory **yellow pages** and **white pages**, **dictionaries**, **calendars** and so forth. As the need for information pull services develops suddenly, they are ideal services to capitalise on the Moment attribute. Pull services such as directory services will be enhanced when the user need is based on location and personalisation.

7.3.6 Find cash machine

For example a service to find a cash machine can be such a service. It can be made better, when it allows you to specify which credit card you want to use, and the system then shows only those cash machines which accept your credit card. Furthermore, in some countries there are families of cash machines, which have different rates. So your card might have no charge on one type of cash machine, but a usage charge in another cash machine. Again the 'find the nearest cash machine' service is better if it can show you all the nearest machines that accept your card, and illustrate which of those you can use without a separate charge.

7.3.7 Comparison shopper

Another example is the comparison shopper. Several comparison shopper services already exist on the fixed Internet, and preliminary tests have been conducted on mobile services. For example the service could tell you which of the neighbourhood gasoline service stations serves the lowest cost fuel. The service would track the prices of the various gasoline stations and keep track of where you are, and based on your location, offer the lowest cost gas station within a reasonably short driving distance, such as within 5 km (3 miles).

7.3.8 News alert service

In addition to the basic news and information coverage provided by the generic news service, there is likely personalised interest in breaking stories in areas which are a particular passion or interest. For example if you work in journalism, you might want access to **breaking headline news**. Or if you work in the financial industry, you might want **breaking financial** and business **stories** sent to you as they arrive, rather than for you to find when you next have time to look at the news. These would need to be very strongly targeted and profiled, and could be offered as a small premium extra service.

Hint to service developers: make addictive services.
As one example, the targeted news alert is a very addictive service if it covers the target person's true interests.

The news alert could be sent as **click-to-read** links to news pages. It would work so that your mobile phone would receive a short headline of a breaking story. You would get the headline for free. If you clicked on the link, you would be brought to a news page with the full story which might be worth 1 cent. The same story would eventually be in the next day's news coverage which you would receive 'for free' as part of your basic news coverage. But having immediate access to it, you would pay 1 cent per news item for only those items which are of really urgent interest. When we consider the cost of the service for the user, seeing the top 20 most important news stories during the day as they emerge, would only cost 20 cents – less than the cost of a regular newspaper. But the news service which provides all of those pages in its next release 'tomorrow's paper' – would get a lot of extra revenue from breaking stories. It is capitalising on the Moment need.

7.3.9 Classified ads on apartments

Personalised classified advertisements can be sent to the user's terminal rather than distributed via newspapers and other mass media. For instance, the user who is looking to buy an apartment can get updated information on available apartments in the neighbourhood of their choice. The service can be made again more beneficial by adding the Movement and Me attributes, by having the system be aware of your location – and allow easy clicking to see available homes in this neighbourhood. When adding the personalisation of the user – looking for a rental flat of two bedrooms in size with a balcony, the service can become very fast and powerful in finding exactly what you are looking for, in exactly the neighbourhood you want. As newspapers and the fixed Internet cannot compete with this type of sorting and personalisation, the mobile Internet classified ads are likely to become soon the preferred channel for such ads.

A good service will of course have **click-to-offer** buttons so you can make an offer on a property, or a **click-to-call** button so you can call the owner immediately. This again builds upon the Momentary interest, the Moment we found the ad, we should call. Not just draw a circle around the ad in the newspaper and think about calling later today.

The content owner for the mobile information services is the major player for this service. The company which has created the electronic maps, hotel and restaurant guides, etc., is currently making its money selling printed versions of that information. The same is true of the newspapers and periodicals. When the momentary need emerges for a mobile user to consume these services, the reason they do so is the availability of information which is seen as valuable. In this case the content is significant, and the better the content – and the more exclusive the ability to access that information – the more the user is willing to pay for it. But in terms of who gets to keep most of the revenues, the content provider, such as CNN news, the *Wall Street Journal* or Lonely Planet Guides, etc., will be able to keep a significant proportion of the money.

Hint to mobile advertisers: mobile services are naturals for sponsorship.
For example McDonald's sponsoring a local town tourist map – showing of course also the locations of the McDonald's restaurants together with the map itself.

7.3.10 Private classified advertising

The good thing about many classified ads, is that they can be user-created content. The network operator should prepare simple forms which are easy to fill in, to allow users to create their own classified ads. The speed would be considerably faster than on newspapers, and the ability to contact the seller would be faster than on the fixed Internet. If you put your old BMW up for sale now on a **used car for sale** mobile Internet site, you might get a call 5 min from now. Remember that the total subscriber numbers for mobile networks totally dwarf those of newspapers, so for example in the UK, Germany or France, there might be 10 million users already connected. Thus if you put an ad for 'Used colour TV for sale' on to the system now, you really might get a call in 5 min. The numbers work in your favour. For the network operator the user created content means that there is no need to share in content ownership. Furthermore, if a simple ad generates 4 min of a voice call to discuss the property, and the owner talks to five people before selling, the amount of extra traffic to the network is considerable for every ad. For this reason the network operator should keep the cost of classified ads extremely low for residential users, such as 1 cent to make a change to an ad (add, edit, remove an ad), and allow free searching of classified ads. The ads should automatically expire, depending on the type of classified ad, anywhere from a week to a month. The network should carefully monitor the classified ads to remove abusive ads.

7.4 Mobile banking

Several of the most basic banking services can already be conducted via a mobile phone today. Scandinavian countries have had several simple SMS (Short Message Service) and WAP (Wireless Application Protocol)-based banking offerings, so that you can check your banking balance, make payments, transfer money between accounts, and other such simple banking transactions. Mobile banking or m-banking has been very successful in the Philippines and in the Czech Republic. In the Czech Republic, one of the major operators, Radiomobil, reports that 10% of its subscribers already use mobile banking. The growth in m-banking is also expected to be very strong. The UMTS Forum has predicted that by 2004 there will be nearly 250 million users of mobile banking around the world.

The biggest hurdle in the widespread use of m-banking is concerns relating to security. We as users must be certain that someone cannot impersonate us and gain access to our banking information or move money out of our account. The security concern is also valid on potential theft or loss of the handset. There are concerns whether the commissioned transaction actually has taken place, but notification services, billing records, and early word of mouth are likely to diminish this worry. There are also of course technical aspects of mobile service security which were discussed in Kaaranen *et al.*'s book *UMTS Networks*.

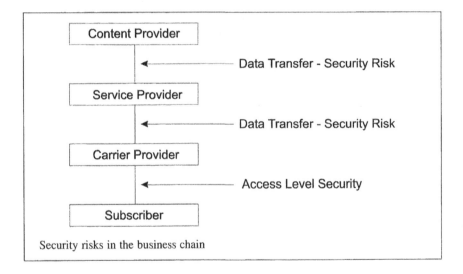

Security risks in the business chain

7.4.1 Access mobile cash from cash machines (ATMs)

One possible new m-banking service is access to a cash machine (ATM: Automatic Teller Machine) and get cash. This might seem odd, thinking that we might walk by a cash machine, point our mobile phone at it, and then be able to get cash. Of course today there is nothing strange to think about putting in a plastic card to access cash from the cash machine. There is nothing to prevent designing a 3G service which allows access via the 3G phone and one's own bank account, to get the money from the machine. It would be no different from authorising a mobile commerce payment from the mobile account to a retailer.

7.4.2 Wiring money to a friend's mobile phone

Another need that often exists is the need to lend money to a friend or relative. Again if it is possible to send an SMS message to the friend, and it is possible for both parties to do bank transactions via mobile phone access to the banking service, then there is nothing technically to prevent direct lending to a friend – in effect making true self-service universal banking possible. It would be much like lending paper money directly from the wallet. Variations of this include parents allocating money to the children's weekly allowance on to their mobile phone accounts, or of parents using mobile banking approval to approve a more costly purchase that the child would like to place on the child's mobile account. Again early examples of these types of services have been seen for example in Germany.

7.4.3 Benefits to users and banks

The utility and access to banking services has gradually evolved over the past 25 years, from visiting the bank teller at the counter, to ATMs often called cash machines, to telephone banking, to Internet banking, and now to mobile banking. The automation in the electronic processing of transactions, and having the user 'self-serve' the financial transactions has produced gradual savings to the banking industry cutting drastically down the need of staffs of tellers and the needs of physical banking locations and offices. For the users the advent of cash machines brought about 24 h access to banking services 7 days a week. Internet banking made it possible to access banking services from the home and office – and during travelling. Now mobile banking brings even more benefits of self-service and digital access providing further savings to the banking world, and even simpler always-connected ability to access banking services for the user, without even needing a costly PC.

The banking world will likely try to port its Internet banking solutions to the mobile environment. Many banks will probably want to try to do this alone and without significant involvement of the mobile operator. There is little benefit from location information, personalisation or co-branding, so the operator cannot bring much in terms of added value to the mobile banking proposition and is likely to keep only a small fraction of the money spent on mobile banking. Furthermore, there are very slim margins and small usage charges related to typical banking transactions, so from the average consumer there is not much money to be made out of providing mobile banking. The benefits arrive more from the savings in automation that are enabled by electronic banking. It should be noted, however, that mobile banking will be a strong enabler of mobile commerce overall, and thus the mobile operators should support the emergence and development of mobile banking even if they do not directly get much shared revenues from mobile banking.

7.5 Mobile games

Mobile games will be such a big, dramatic, visible and revolutionary business that a whole book could be written just about mobile games. The electronic gaming industry features names such as Nintendo, Sony Playstation, Sega and Electronic Arts. The industry is the fastest growing segment of the entertainment industry by a wide margin, and in volume of sales has reached the size of the Hollywood motion picture industry box office revenues. It is reported that the games industry had reached 20 billion dollars worldwide annually, growing at annual rates of about 17.5%.

The games exist in incredible variety, from trivia quiz games to sports and racing games to role-playing games; from space games to fishing games; from animals, heroes, popular characters to shoot-em-ups. Games are also often built around popular **game shows** such as *Who Wants to Be a Millionaire*. The games are geared towards the young and youth-oriented segments and thus they play on fantasies and allow a lot of creativity and imagination. Of course early SMS, WAP and I-Mode based games are still relatively simple and unsophisticated, but in 3G these will evolve into much more complex and realistic games. Already in Japan there is a multi-player **role-playing** Samurai game which allows half a million simultaneous users to participate in the game.

7.5.1 Try it for free

With the PC-based games the industry learned to take advantage of the 'first game free' model where subsequent game, levels or difficulty is made to be sellable. This has been a key to generating revenues from games which often have no previous users or name recognition. Variations include the free trial, the free basic levels, or a limited amount of time that the game is free. These kinds of free trials should also be utilised in 3G to introduce new games to the mobile gaming marketplace.

The gaming industry has pioneered revolutionary bundling, cross-advertising and sponsoring methods, including putting games on other devices where they did not exist before, such as the **Snake Game** in Nokia phones, or bundling the solitaire and minesweeper games into Microsoft Windows. Various game-oriented motion pictures have emerged such as *Tomb Raider* and *Super Mario Brothers* which were designed to generate more revenues from the game of the same name.

The industry is driven by young eager and devoted programmers who want to create the ultimate games. The industry is known for its innovation, and also its very fast reaction ability to change and opportunity. Games also wear out fast, so few electronic games remain popular for long. The industry has to innovate to survive.

Those developers have been looking for more resolution on the screens, more portability on the terminals, more interactivity and ideally connectivity and multi-player remote gaming possibilities. The gaming industry is also very shrewd in understanding the money side of the business, and very many gaming companies are very profitable. Now they have a new platform which combines all they want, and the gaming industry is ready for it. 3G is the perfect environment for games.

The 3G terminal is the ideal gaming platform. It has a high resolution colour screen. The terminal itself is pocketable and small. The 3G terminal is going to reach penetrations that are much larger than the whole PC penetration or any of the current gaming consoles such as the Sony Playstation. Most importantly, potential gamers will be carrying the 3G phone with them at all times. The 3G network can deliver a lot of capacity and allow game updates, remote gaming, multi-player gaming and a built-in interface to handle money and billing. And the 3G terminal will be always on, so interactive games can have real-time updates to scores, etc.

3G terminals themselves will have expansion memory to allow various game elements, graphics, and the gaming engine to be stored on the 3G mobile phone. This allows a game programmer to build the game so, that only changes in the ongoing game are transmitted, while the mechanics of the game are downloaded once and stored on the phone. As 3G terminals become more advanced in the future, their memory and storage ability will grow fast, creating opportunities for more advanced games to be created, downloaded,

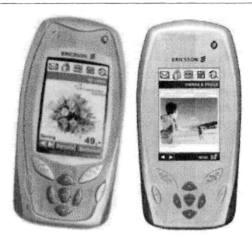

Imaging will play a large part of UMTS phones and the majority of terminals are expected to eventually incorporate large full colour screens and cameras

stored, and played. This is also bringing a need for upgrades to phones, both in terms of memory chip and storage device upgrades, as well as upgrades in the gaming 3G terminals themselves.

7.5.2 Games are addictive

Games are found to be particularly addictive to their users. As many will attest, even the simplest games such as the solitaire game on the PC – or such games as Tetris can create a strong addiction, at least over the short term. The most fanatic gamers do not care about the time, place nor cost. For the 3G operator, game players are an ideal target audience. They are already accustomed to similar devices – gaming consoles tend to have small screens, small control buttons, and various menu-driven instructions. Game players have yearned for multi-player options, to play with friends together, or to play with friends or strangers remotely. As anecdotal evidence suggests that for children – boys and girls – when parents ask what they want on a trip to visit relatives, the child will say the most important three things to take along are the mobile phone (number one), the Playstation/Nintendo/equivalent (number two) and the Walkman/CD player (number three). In the 3G environment, the youth-oriented 3G phone will be all three in one device. It will be the mobile phone, the music player, and the gaming device.

7.5.3 Business uses of gaming/simulations

Businesses can also use gaming, and many uses already exist. Professionals do not like to call it gaming, as that has associations with childhood, so business gamers have adopted the term 'simulations' from military **war gaming**. Simulations are already used in various business instances from the training of airline pilots to testing engineering solutions. Gaming/simulations can be used as a training and testing vehicle in most industries. For example monotonous work quality can be improved with simulations, such as currently airport security systems can insert the image of a hidden gun into the X-ray equipment visual display, so that the guard who checks tens of thousands of bags per day will still be alert when something sinister goes by. Some very advanced gaming simulations exist to mimic the competitive environment, such as Nokia's 'Equilibrium' simulation and workshop to train telecoms operators to prepare for telecoms competition in 3G and beyond. Such 'war gaming' simulations are the only way to prepare for upcoming competition, and such simulations also allow for testing the effect of action in given scenarios without revealing one's plans and strategies.

3G provides a powerful platform for simulations. The terminal itself can be a gaming tool, and as the world becomes ever more digital, it means that most of the 'materials' that might be needed for a simulation are already in existence. Here the content and the innovation in building the simulation are driving the value to the user, and thus the network operator's share would be limited to the data delivery and the value-add of personal information, and, depending on the nature of the simulation, possibly location. Simulation is likely to be a strongly growing area of industrial and business training in the near future.

7.6 Real services today on the Moment attribute

The ability to offer services which address the momentary needs of mobile phone users is not a futuristic vision. It is happening today around the world. Numerous simple such services already exist as this book is going to print in the spring of 2002. It would be impossible to list all, but a few illustrative examples are listed here to prove the overall concepts and that we are not in deed talking about a far-fetched future.

7.6.1 Existing mobile information services

Information services have emerged already on SMS, WAP and the Japanese mobile services. News services have also experimented with totally free content, for example in Finland and the UK already during 2000, where SMS-based news information was sponsored by advertising.

7.6.2 Apartment locator

The first examples of a real estate agent providing information on available homes and based on mobile information have been trialled in Finland with the country's biggest real estate agent, Huoneistokeskus. The first version of the service required that the mobile phone user types in the street address, but the service was being developed to automatically recognise the location and provide available apartment information based on mobile phone location.

7.6.3 Soap opera updates

The Philippines have built many mobile Internet services on to the SMS platform and one of the more popular ones is the updates and related services to the soap operas running on TV.

7.6.4 Mobile cartoon

Cartoons have already started to appear on mobile services. The most famous of the early cartoon services is probably the Bandai characters in Japan on DoCoMo's I-Mode service. The principle behind Bandai is that there is a daily update to the cartoon characters.

7.6.5 Mobile banking today

Banking has also been migrating to the early mobile services. Of course many banking services can be used as voice-driven or phone keypad-driven systems as telephone banking, and those can be accessed from fixed or mobile phones. Some of the more advanced mobile banking solutions were first deployed in Finland and Sweden from 1999, but during 2000–2001 the big development in mobile banking has been in the Philippines where a service called Smart Money offers a wide range of mobile banking services; the service has won international awards in innovation.

7.6.6 m-Lottery

Several countries with national lottery boards or companies are exploring the opportunity to play the lottery via mobile phone. An early example of it being possible comes from Finland where playing the lottery via mobile phone has been possible since 2001.

7.6.7 Transferring money to friend's phone

In Germany it is possible to 'wire' money from one mobile phone to another. People use it to lend each other money and to take small payments. A similar ability exists in the Philippines.

7.6.8 Mobile game adoptions of familiar games

Mobile games are another area which has a long history and development happening all around the world. Real services around trivia game ideas, such as Who Wants to Be a Millionaire have been launched in many countries.

7.6.9 Mobile network-based games

Games involving characters and personality are evolving fast in Japan, where the best example during 2001 is the Sumo Wrestling Game, where players have their mobile wrestling characters play against others. But in the Sumo Wrestling Game it is not enough that you are an astute player of the game when you play it, you also have to train and feed your own Sumo Wrestler throughout the previous days to prepare for the match. This game is in essence a mobile phone update of the Tamagotchi phenomenon from a few years ago.

7.6.10 Virtual date game

Another gaming example are the games where the player interacts with a virtual 'date' and games like these exist already in Hong Kong and Japan. Typically for example a male player sends messages, flowers, etc. to the date and she responds warmly or coldly depending on how long it was since the last contact, and what was the current show of affection, etc.

7.7 Last moment on Moment

This chapter has examined how money can be made using the Moment aspect of 3G services. The other aspects of the 5 M's are of course important as well, and none of the services discussed in this chapter relies on the Moment attribute alone. Trying to use time, save time, move time (re-schedule), multi-task, and manipulate time in any way we can is an ever increasing need by us all. Services which help in doing so will bring value to the user. That value can be translated into revenues and profits to the various players involved. Remember that while all 5 M's can be used to enhance digital content, the Moment attribute is the most powerful agent to action. When we get a Momentary urge to do something, cost is often ignored as an issue at the impulse. Make it very easy to react to sudden, Momentary needs or wants.

The future will be here soon and in that future man will be ever more stressed and concerned about time, finding more of it, making the best use of it, and often being frustrated at the speed of time going by. For those of us working in getting 3G services up and ready to make money, it will always seem like time is running out. At least now with new mobile services, it seems that everybody is in the same boat and everybody is complaining about being stressed to keep to schedules. Don't wait to introduce the perfect service, remember that even more than before, in 3G you are in a hurry. As Benjamin Franklin said, "Remember that time is money."

8

'My closest relation is myself.'
Terence

The Profits of 'Me' Services:

Extending Me and my community

The mobile Internet's ability to provide personalised content and to extend the self into the near community will be highly significant to 3G services. The Me attribute is the one we hold most valuable on mobile services and are willing to pay for services to be tailored just to us. The Me attribute includes personalised content, exclusively relevant content, customisable and modifiable content; and the ability to interconnect and share with one's community. The community could be family, work colleagues, friends, people sharing the same hobby, etc. The Me attribute also includes the communication in one's own language, and to allow multi-session use, i.e. while viewing a promotion for a movie and to also talk on the phone about that movie. The more a service provider can differentiate any given service to every customer, so that they can observe that what they receive is different – and better to them – than what their friends and colleagues receive, the more they are willing to pay for the services.

The services described in this chapter have a high benefit on the Me attribute. Of course as with all mobile services, any such service can also benefit from the other 5 M's, but those in this chapter are particularly strong on the Me attribute.

The Me attribute is the most personal of the 5 M's. It is therefore the most meaningful, and also the one to which people assign most personal value. The

Me attribute is also why we view the mobile phone as the most personal of our gadgets, as was discussed earlier in this book. The Me attribute has a lot of aspects which relate to the ego and thus may feel 'silly' or trivial. Do not underestimate the Me attribute: it is not only the most valued of the 5 M's and is readily billable, it is also the biggest key to customer loyalty. The needs of the user are least well understood by mobile operators of all the 5 M's, and customer intelligence, segmentation, CRM (Customer Relationship Management) and other such methods should be used to learn what it is that any given customer really wants and values.

8.1 Rich calls

Rich calls are voice calls which allow for further enhancement of the call, typically with images or sounds or both. There are many ways that calls can be enhanced with images, clips, chat and whiteboard. The most obvious use is one we can all identify with, whenever we are lost and speak to someone trying to give instructions.

8.1.1 Draw Me the Map/Draw It for Me/shared whiteboard

Draw Me the Map is technically called a 'shared whiteboard' rich call service. Shared whiteboard – a typically cryptic techno-babble term – may sound strange, but a simple example will illustrate it. If you and I were talking on the phone, and you were lost asking for directions, wouldn't it be nice if I could draw the picture at my end, and you could see it, as I draw it, on your end. This is a shared whiteboard. That is why Draw Me the Map is more obvious in describing the benefit to the user.

> **Hint to 3G network operator: don't use technical terms in services,** such as 3G whiteboard. 'Draw Me the Map' or 'Draw It for Me' is a much more descriptive name which non-technical users will understand.

There will be numerous other uses for the service, where it is perhaps best called 'Draw It for Me', for example describing patterns, diagrams, shapes, etc. The sending person draws on the mobile phone with for example a stylus pen, and the receiving person will see it as the drawing takes form. So in our lost example, if I tell you where you need to go, and draw you a map, and explain while I draw, you will have a much better understanding of what I mean. Best of all, if you have questions, you can point to the drawing at your end and I will see what you mean and be able to answer. Draw It for Me will be a new way of giving guidance in the near future.

This service illustrates the richness of rich call services. The traditional ultimate rich call service is the shared video image – which can be of use while we talk. The difference between rich call video, and video call video, is that in video call the image must be perfectly synchronised, to a fraction of a second. Otherwise the lip synch of speech is distorted, and the conversation focus goes into trying to understand and we focus simply on why the lips don't synchronise with the speech. It is the same annoying pattern sometimes seen in old movies when the synchronisation of the sound is somehow off. Video calling is expensive because the network has to designate a lot of bandwidth to guarantee constant real-time synchronisation of the image.

8.1.2 Shared Video Image

In Shared Video Image there can be 'live' video image which is slightly delayed. This would include a lot of uses, but one will be the initial 'facial hello' of greeting friends, while not connecting via video call. The caller can turn on the video camera, take a 5 s snapshot of the caller just smiling, and send that while the conversation is going on, to show the new haircut or just to give a smiling image to the other caller. The shared video image can of course be of any happening and of any duration. 3G allows for quality of service classes, by which this type of 'near real-time video' to be offered at much lower cost than video calling. The shared video image can be used to build friendship and community, and is likely to become quite popular as a greeting – show me what you are wearing today, etc.

8.1.3 Games and rich call

A huge growth opportunity is mixing voice calls and mobile games. At the one extreme there are pure voice call occasions, which can be enriched with showing scores. At the other extreme are pure games which get richer through the ability to talk to the opponent. Various 'talk to' and 'broadcast' buttons and features can be programmed into a game, for example in a paired game like chess to talk to your counterpart. Or in a group network game like a battle game to talk to your opponent, or to your partner on your side of the team. Furthermore, in games of group chase the chased 'prey' (or the game itself) could broadcast to all chasing, etc. The introduction of machines talking, and sound effects, are a huge additional enhancement to any gaming experience. And the ability to pipe simple background sounds is yet another dimension that game developers can use to enrich the gaming experience and build gaming atmosphere through rich call.

8.1.4 Original music and rich call

There already are mobile phones which can be used to compose ring tones and already some handsets exist with MIDI (Musical Instrument Digital Interface) abilities. In other words a musician could use a mobile phone as the instrument to record the MIDI soundtrack of a melody, and then use a digital synthesiser to modify that soundtrack to make it sound exactly like a saxophone or electric guitar or violin or drum set, etc. This means that amateur musicians – or even professionals when feeling creative – could turn an ordinary phone call into an opportunity to play music to the other caller. As long as convenient music software is built into the handsets, or provided via the network, this could mean that a musically talented Romeo could serenade his Juliet via the mobile phone.

Rich calls are communication. There is very little content that needs to be, or in fact can be provided by the network operator or its partners. Rich calls will move content, but that content will for the most part be user-created content. As such it is particularly good for the network operator, as the services can be priced to be profitable, and no money needs to be shared with those who create the content – the individual callers.

8.2 Personalising calls

In 3G we will have the possibility to personalise our phone calls. There will be many ways ranging from selecting how the receiving phone will ring to what picture it will display. The pictures we display will lead calling towards video calls, without being video calls. In fact there is likely to be a lot of confusion with some users and even mobile phone salespeople in explaining the nuances between various picture delivery possibilities. Some personalisation of the phone call can be done even without the image as we will see next.

8.2.1 Personalised contacting

We know already about mobile phone ring tones that allow us to personalise our own phones so that my phone sounds distinctly different from your phone. And already most phones allow setting up calling groups, etc. so that when your wife calls you, the phone sounds different from when your work colleagues call you. But a new technical feature called SIP (Session Initiation Protocol) allows for much greater personalisation of the call which you initiate.

For example in 3G you can select your own ring tone which is sent to the person you are calling. In this way if you want to be dramatic, you could send

8 Vignettes from a 3G Future: Nearest Parking Place

The navigation and intellect in my car is getting ever better. Now the car is able to learn how I like to park and checks with the parking lots at work, shopping centres and sporting stadiums, etc., to tell me where is the nearest empty parking place for me. I love it, no longer do I have to cruise the parking lot up and down the aisles thinking do I want to park this far, or will I not find a place closer to the entrance. In places where there is a parking fee, the fee is also paid automatically and billed to my phone account, so I never need to pick up slips of paper and pay at parking meters.

The integration of the parking lot intelligence which can detect which parking places are in use, combined with the navigation systems of cars will take some time and trial and error. But intelligent parking lots are likely to become features of premier shopping malls and eventually become commonplace for most places where parking is an issue. The new automobile fleets will soon be intelligent enough and have the ability to connect to such systems, and the interest of the car manufacturers is to find clever uses for their new technology.

some Beethoven's fifth, or if you are a James Bond fan like me, perhaps the 007 theme, or when missing your loved one, your selected ring tone could be Stevie Wonder's 'I Just Called To Say I Love You'. The possibilities are endless, and of course the selected sending ring tone would further allow us to indicate our personality in the way we communicate, just like a few decades ago people selected coloured stationary on special papers when writing personal letters. The operator and some content providers and application developers can build libraries of sounds and images that could be used for such cases. The business case would be very similar to the ring tone and downloadable logo business in second generation cellular networks.

8.2.2 See who is calling/Calling Line Picture

One of the exciting yet simple new features of 3G services will be the ability to transmit the picture of the caller. Similar to the way currently the fixed and mobile phone networks transmit the calling line identifier (phone number) and based upon that you can program your telephone to show the name of the caller, soon it will be possible to transmit a simple picture of the caller. Then you can see who is calling. Several services can be built around this feature. Calling Line Picture will not be video calling, but will bring video calls a lot closer to comfort, as the face of the caller will be presented when the phone rings. Initially it will seem like a novelty, but soon it will also be a tremendous convenience, where you don't have to think about mysterious phone numbers but can immediately and very literally see who is calling.

The data capacity requirements to deliver a still image of a caller are not remarkably great, the image could be delivered in almost the instant that the phone connection would be made. The service would need the option to turn the sending image off, and of course convenient ways to update the calling line picture. There should be the ability to switch images, for example, if different family members call from the same phone.

> **Hint to 3G network operator: the Video Call is distant future, start the change in behaviour with sending images.** Services around the Calling Line Presentation and Face Catalogue will get users accustomed to like to see the face of who is calling, make sure it is easy to adopt and use.

8.2.3 VPN with Pictures/Corporate Face Catalogue

Calling Line Picture will enable a 3G version of the 'Face Catalogue' or corporate directory with pictures. Early on, corporate customers are likely to be among the first to have significant numbers of 3G phones, and thus they are likely to be first to use Calling Line Picture. The Face Catalogue service could be built into the options of the corporate customer service called VPN (Virtual Private Network), which allows dialling short-number extensions even if offices or mobile phones are far away. Obviously in a company of a dozen employees a face gallery is of no use as everybody knows everybody, but in a multinational company of tens of thousands of employees, the ability to put a face with the calling person's name would be a great asset and improve overall communications.

The mobile operator providing the Face Catalogue service would want to allow easy photo generation and central storage. The system could easily be enhanced by showing also the caller's name. A creative innovation would allow links to calendar software, e-mail, etc., so that you could enhance the virtual digital work space being used, so that something which starts as a phone call, could soon be expanded into a shared whiteboard simply by making simple clicks and selections. The integration and central database functions of setting up such corporate solutions would be natural extensions of a more sophisticated communication solution built around the VPN and centralised scheduler.

> **Hint to 3G network operator: VPNs are Sticky Apps for corporates.** If your target audience includes the large corporations and multinationals, make sure you build your VPN to have strong features and hook your corporate customers with VPN.

8.2.4 Look Whose Calling/Private Face Catalogue

The initial Face Catalogue feature with business use is likely to expand further, so that soon the same service would be available, on a permission basis, to all callers. Some will want to remain anonymous, but others will enjoy having their face displayed whenever they make calls. As it would not be a live picture, there would be no worry about the hair and makeup being wrong, when one has a 'bad hair day', etc. The primary costs involved to the 3G network operator is the file storage space. Still the user is likely to be willing to pay a small monthly fee or an initial set-up fee to have this feature turned on.

8.2.5 Personal Publicity Photo

A variation of the calling line picture is the ability to download your own picture, and to send it to others. In many industries and in personal life, it would often be convenient to be able to send a current colour picture of yourself. With 3G and your 'face' already stored on the network, this becomes a natural extension of the face gallery feature. The 3G network operator could also entice its users to visit the company stores or those of its authorised dealers to get digital pictures taken and stored on to the network. In this case, the 3G operator could store two versions, one the low resolution small file size image for the Face Catalogue, and the higher resolution image for personal publicity photo usage. As digital cameras become more popular, and integrated with the 3G phones, soon most pictures would be taken and updated on 3G phones.

8.2.6 Lets Talk About This/Image of Our Topic

Another variety of the sending of the image as the phone starts to ring, is to send an image of the topic you'd like to talk about. So for example if you are a Bayern Munchen soccer fan in Germany, calling your football friend, you might send a picture of the star player scoring a goal. In this way when your friend sees the phone ringing, the image of the goal being scored tells him immediately that the topic will be yesterday's game, and likely will guess it is you calling. The Topic Image feature will probably grow with experimentation and find its own users and uses, especially with younger people. Creative people will innovate using built-in digital cameras, and images lifted from news services and popular sites to create personalised greetings, much like today people create personalised wallpapers on Windows.

8.3 Show Me

Show Me is another rich call type, but likely to be one of the most popular new services with thousands of uses which warrants its own discussion. Once when callers learn to use the ability that they have a live video camera on their phones, they will start to use it to literally show the other caller whatever it is that they are talking about. And soon after that, the callers will start to think at the next level, so whenever they hear something they don't immediately understand or know, they will ask the other caller to 'Show Me'. The possibilities are endless and this part of this chapter will only scratch the surface.

8.3.1 Show Me Personal/Remote Eye

A happy new father could show the relatives faraway what the first born baby looks like. A traveller could show panoramic views of the sights of the faraway location. If father happens to be away on a business trip when junior plays in a big game, mother could show a key play of the game live, via Show Me. The youth could be inviting guests to a 'happening' party by showing how many people are there and how live it is. The husband could call the wife and show the options at the store, and have the wife decide which colour shirt goes with his suit. The motorist with engine trouble could use Show Me. The 8-year-old girl could show all friends the new puppy. As men don't ask for directions, perhaps showing me the view from the car could be an answer to not admitting being lost.

The utility of this remote 'eye' is that the image does not need to be live for long – it can be stored at the receiving end, so the other person can review the recorded image if a longer view is needed. And the remote person can direct the camera end to show a close-up, or show another option, or the view from the other side, etc.

8.3.2 Show Me Repairman

The same show me idea will work in numerous business uses, for example the PC repairman might be able to ask the caller to switch on the camera and show what the PC is doing that is causing the problem. The PC repairman might immediately see what the caller would never think of explaining, and in this way deliver very much faster and better customer service by using a 'remote eye' to the site.

8.3.3 Show Me in Other Business

Again there are too many examples to ever think of them all. A real estate agent could show the new apartment to a prospective buyer via show me to entice the buyer to come and see the place in person. The images could be stored at the real estate company's site and subsequent interested parties could do virtual visits to see the property. A sales representative could show a sampling of the new products. A lost luggage search could be faster 'showing me' the image of the found luggage. An engineer could ask a technician to show me the actual problem that the technician has uncovered, this could work with architects and constructors, designers and drafters, etc. In any emergency service cross-responsibility situation, for example a police-man at an injured person (where an ambulance would be needed), etc., the emergency specialist can probably assist very well, if given guidance by a professional, and that professional is given eyes. So the policeman could call the emergency doctor on call, who would ask the policeman to show me. As with rich call services, Show Me communication is user-created content, and there would generally be no need for the network operator to share content revenues with partners.

8.4 Video calls

Video calls have been the ultimate science fiction promise of telephones of the future. Video calls have also been promised to be 'just around the corner' and about to take off, for decades. It is perfectly possible to have video calls today, with many video calling telephones on the fixed networks and for example on ISDN (Integrated Services Digital Network) but the service has not taken off.

There are many theories why video calls remain on the expectation list but never seem to make it to real mass-market services. 3G will allow having video calls on the mobile phone. And most 3G phones will be able to receive video calls, and many will be able to send video images to create video calls.

I do not want to spend much time speculating on what will happen. But the rich calls and Show Me services described in this book will create an environment where it will be common to send and receive moving images. The face gallery and personalised greetings will by their part reduce the barriers to showing one's face when calling. From there it is not a drastic leap in usage to incorporate live video on to the call. Video calls will be more popular in the 3G environment than today, after a reasonable penetration of video-sending handsets – i.e. phones with built-in video cameras. But how commonplace will video calls become, only time will tell. Definitely there will be a lot of early debate about are video calls a good development or not for society and communication, etc.

8.5 CRM – Customer Relationship Management

When one thinks about the Me attribute and business, it might be easy to conclude that there are not Me services for businesses as Me is myself, the reader of this book. Actually, if we consider it from the business entity's point of view, then of course the business also wants Me services tailored for the *business*. The business wants services to be personalised to the business itself ('Me') and to build the community for the business. From the area of the Me attribute, but serving primarily exclusively business needs, the best example

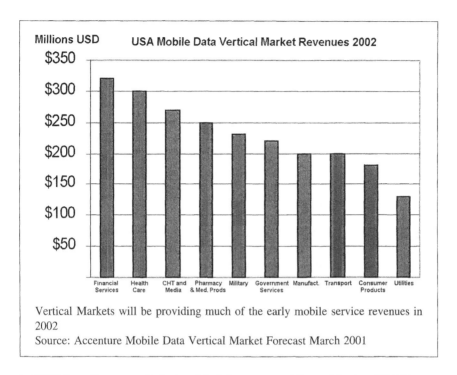

Vertical Markets will be providing much of the early mobile service revenues in 2002
Source: Accenture Mobile Data Vertical Market Forecast March 2001

is 'CRM' or Customer Relationship Management. For achieving CRM benefits to its customers, the operator will need to examine vertical markets.

Numerous systems and solutions exist to address CRM needs, and many consultancies make a lot of money implementing them and training various customer-related personnel to make good use of the CRM solutions. CRM system benefits can be enhanced by mobility, personalisation and time sensitivity. A thorough mobile version of the CRM needs to keep track of where you are in relation to the customer – i.e. if you are in a city where they have major installations, etc. The system should keep track of which of your company representatives have met with the customer recently, and what has happened as result of those meetings. The system should track your customer's specific developments – such as orders placed, complaints logged, and press coverage of that customer. You should have them in real time, and any breaking news should be pushed to you.

A CRM system will depend on the type of business – an on-line wedding services store will have varied customers who rarely return, whereas a company servicing jet engines for passenger jets will have a very defined set of the same few airline customers year after year. But both can develop their business by understanding their customers, and building that relationship. Almost any part of the CRM could be enhanced by adding a 3G mobile interface. Collecting data about the customer and relationship, i.e. customer information, what has a service technician seen when he visited their site, what has the press talked about, what does the Account Manager think, what do they say officially, etc. Any and all of these could be 'entered' via a mobile terminal. The convenience of having immediate, always-on connection to the CRM is that all of those people who do not naturally think of entering the data will have natural barriers to bothering to do so once they return to the office, or once they have the free time.

8.5.1 CRM money rewards

A creatively motivating business can utilise the micro-payment functions of the 3G network and reward its data entry contributors with small payments building up to their paycheques. This could be real money, or if the company has products that the employees usually like to buy, the money could be 'company money' which could only be used to buy company products. That could include products from the company's gift shop, etc., if one exists. The point is that for the first employee who brought in the information that a competitor's product has been seen at the customer, the system would reward with a small additional payment to the paycheque.

As small payments are no problem for the 3G network, such rewards could be tiny for really small bits of relatively unimportant information. This could be for example entering the new title of a customer's executive, who otherwise already exists in the CRM customer executives listing. This kind of small contribution could be worth pennies. Then, entering a new person's name and phone number could be worth dollars, and so forth. At the other extreme finding information about a competitor's bid to our customer could be a substantial amount, for example 0.1% of the actual value of the competitor's contract offer. This, depending on the size of the contracts in that business, could range from dollars to tens of thousands of dollars and more.

Using cash to reward employees for contributing to information beyond their regular duties, is nothing new, but its suitability would vary by company culture and the motivational systems used in that given country's business environment. What 3G and an integrated CRM system bring, is the ability to make the tiniest amounts of information also payable, without crippling bureaucracy of filling forms and approving with human resource departments and payroll accounting.

8.5.2 CRM Location Lock

Of course the customer will be curious to find out what you know about them, and what you store about them in your CRM database. One of the exceptional benefits of putting the whole CRM on 3G is the power of the Location Lock. The Location Lock assigns location requirements for some type of information so that specified information can only be accessed when the access device is in a given location. This could vary easily by your company offices, that the headquarters might be the only place where the legal contracts can be accessed electronically. Then, like in the world before CRM, if anybody wanted to take the information to a meeting with the customer, they would have to print out the contract and discuss it 'over paper'.

So for example, some very sensitive strategic information, and definitely any embarrassing information, about the customer could be locked so that it can only be accessed from *your company* office locations. The sales rep or service technician could safely show the CRM entry for that given customer and the system would not show the 'hidden' entries behind a Location Lock. As an example, someone at your company might hear a rumour that one of the major decision-makers is interviewing for another job with some other company. That may be critical for your company to know. Equally it could be damaging to admit that you know, especially if such a rumour ends up not being true.

8.5.3 Dynamic Location Lock

Location Lock can be built of course to be dynamic as well, by which defined people would have control over when and where the limits of the Location Lock would be enforced. For example the system could be so aware of the Movement attribute, that the Location Lock would extend to any place your CEO travels as a travelling proximity. In this way the vice presidents and experts who happen to go to a strategy retreat with the CEO would have full access to the information otherwise locked to their HQ location.

8.5.4 All aspects of CRM can benefit from 3G

There is much more to CRM of course, and volumes have been written about the subject. Those consultants who work in CRM should explore this book and see how any of the services described here can be used to make the CRM better for any given customer. For example the ability to store 'face galleries' of the customer's personnel, and innovative ways to capture the images and assign names to faces. Another example is a network storage solution for Show Me clips, where a library of Show Me video clips relating to your company servicing that customer could be collected, building eventually a very detailed and complex service history. Here a picture could be worth much more than a thousand words, and a moving annotated Show Me video clip of exactly how that customer's solution has been built could be worth much more than a thousand pictures.

CRM design, system integration, and optimisation is typically a long process re-engineering task in any business and mostly involves specialised CRM experts and consultants, as well as IT specialists. Those network operators who want to offer leadership in 3G-based CRM systems and extensions will need to make a concentrated effort into understanding their business customers, and typically would need management consultant partners as well as IT integration partners. If the operator is closely participating in the design of the CRM solutions, it would have a valid claim to some of the revenues. If the network operator leaves the design and integration to third party specialists, then the network operator would be reduced to delivering the data bits of the solution but little else in terms of revenues.

8.6 Messaging and love life

The 3G environment will have a lot of value-add services, but still the primary reason we use a phone in the first place is to communicate and be connected. The mobile calling trends and patterns have been growing and evolving for over a decade and are well documented. But the SMS (Short Message Service) text messaging culture is still quite young, with the first SMS service turned on only in 1994 in Finland. And even then, initially SMS was not seen as, nor expected to become a method of communicating by the masses.

8.6.1 Dating and messaging

The early adopters of SMS as a communication and community tool were of course young people all around the world. There are only a few societal studies so far on how SMS has changed society, especially among the youth, even though there is a lot of anecdotal evidence. An early and very insightful look into how SMS has changed society is Timo Kopomaa's book *The City in Your Pocket*. Among the many findings Kopomaa established that

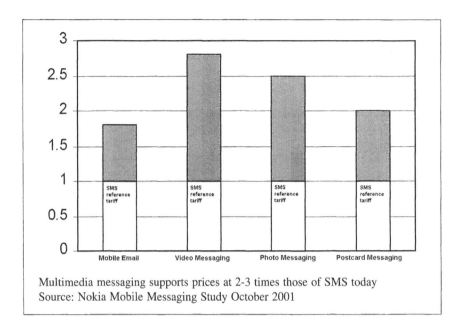

Multimedia messaging supports prices at 2-3 times those of SMS today
Source: Nokia Mobile Messaging Study October 2001

young people prefer to start dating via SMS. SMS is seen to be less painful and a more flexible way to ask someone out – and to handle rejection – and this feeling is shared by both boys and girls.

The sending and receiving of text messages is seen as evidence of affection, and boys and girls expect to receive messages from their loved ones, at least one per day. Young men and women send each other love notes, poems, personal jokes, messages with innuendo and slang terms and cryptic messages. They are a new form of flirting and of bonding. From a 3G services theory point of view, there is hardly anything more personal than the private expression of affection to another person. So this is a Me attribute service at its very heart.

8.6.2 SMS text messaging

The first significant lesson to be learned is that SMS text messaging is a new form of communication, where old rules do not apply, and which does not inherently detract from other established means of communication – even young people who are totally hooked on SMS, still call each other on mobile phones, and send e-mails. But SMS has arrived as a new phenomenon to add to the communication traffic.

The second lesson is that SMS text messaging is a learned habit. And it is hard to break, one could argue that it is addictive. Certainly anecdotal evidence suggests that people accustomed to SMS are not willing to live without it. So cellular network operators should embrace SMS text messaging, deploy it, and entice all subscribers to try it. They will get hooked.

Hint to 3G network operator: get all of your customers hooked on SMS now.
All evidence shows that SMS is a learned habit and becomes very addictive. Get the non-users to try it now and become regular users of it. Use occasional free SMS promotions, competitions and promote the forwarding of messages.

As with most user-created content, the network operator gets no extra revenue from this traffic, but then again, SMS text messaging at about 15 cents per message is very profitable for the network operators.

8.6.3 Dating services

It did not take long for commercial enterprises to notice that young people were using SMS text messaging to start the rituals of finding romantic partners. Soon various chat rooms, dating servers, personals ad systems and attraction servers emerged. During 2001 they were still predominantly SMS based, but the ability to add pictures and sounds make such dating services naturals for evolution into Multimedia Messaging. With dating services there are gimmicks and virtual communities which serve to attract ads or messages or communication traffic. The gimmick could be the ability to appear with an alias or anonymously, or that the virtual event is sponsored by a local nightclub, or is a mobile version of a popular Internet or newspaper-based personal ads service.

8.6.4 Multimedia messaging in dating

The advent of multimedia messaging will bring new opportunities for creativity. Now the young Romeo can send his Juliet drawings, pictures, music, and other such 'artistic' output, in addition to the love poems and love letters, etc. which are possible in the text-only medium of SMS text messaging. Young people are not inhibited by limits of one technology and tend to be quite creative. So they are perfectly poised to make the early masterpieces of artistic application of multimedia in messaging. Multimedia messaging, like most messaging, is again person-to-person communication, with user-created content. Thus there would be no revenue sharing with content providers.

8.7 Communities

One of the most powerful of the Me attributes is the community nature of defining oneself by those that one includes and excludes into one's own community, as well as the acceptance (or rejection or ejection) into communities. Communities are unofficial groupings of people who communicate with each other. Typically they centre around hobbies and interests, and at work informal communities often naturally form around teams and groups of colleagues. Communities form around anything such as a joke distribution list or a competitor intelligence mailing list or a chat room of fans of Mariah Carey, or the collection of parents who have to arrange car pooling for their daughters' ballet practices.

8.7.1 Spreading messages to the community

People in communities use communication to share information but also to build and confirm the commitment to the community. Communities use any powerful communication media, such as e-mail and recently increasingly SMS, even mobile chat. The power of breaking into a community is that communities tend to strengthen messages which have been accepted by gate-keepers into the community. So if you get one of the influential parents in the ballet daughters' parents' car pooling group to think a given dance shoe is particularly good, the parent is eager to promote it to all in the community. And an endorsement by someone in your community is likely to be taken more seriously than a commercial endorsement for example by a celebrity in an advertisement. Often the community will use viral marketing to transmit commercial messages to each other. Viral marketing is discussed in Chapter 9 on Money services.

8.7.2 Multiple concurrent communities

What makes community services even more of a 'snow-balling' effect is that ever increasingly those who are in one community, are also in other communities with some overlap. So the ballet girls' parents may have sons in football practice. Of course not all, as some don't have sons, etc., but this would mean that perhaps half of the mothers taking daughters to ballet class, also meet each other bringing sons to football practice.

The messages that community leaders could send to the community are not necessarily limited to the strict topic of what the community is about. People often have the ability to stretch the definition of what all might be seen as common. So for example if a pair of parents enjoy a particular movie, they could easily 'market' that to the other parents in the community. An astute marketer should look into tapping this power, and even consider arranging samples and free trials for some community opinion-leaders.

8.7.3 Services for community-building

Apart from using the normal person-to-person communication tools such as voice calls, e-mails, SMS text messages, mobile chat, etc., communities develop their own needs which are similar to small businesses. There is often a need for a community bulletin board. There is a need to manage phone numbers, addresses and schedules. The groups may have little budgets to manage, trips to arrange, etc. The community may be a significantly large group and thus of interest to a travel agent, etc., and the group could organise so as to get some bulk purchasing benefits.

The 3G network operator should try to make this type of community-setting a very easy activity for any of its subscribers, so that the community could have an identity all by itself, with identified 'sponsors' of the community from the subscriber base. For example many business services could be offered to the community on a pre-paid basis. For the individuals setting up say a stamp collecting club or a skateboarding group or a fashion photography enthusiasts club, the chance to get the community 'officially' recognised by the telecoms operator would lend it a lot of authority, further enhancing the appeal of the community to its members. Members could participate in the community by paying their share of the pre-paid account to become 'official' members.

> **Hint for service developers: passion is a key to successful service.** Find something that some customers are passionate about, and build a service around it. For example look at Manchester United football club or skateboarding youngsters.

Separately 3G operators should publicise the innovations that the informal communities implement. Some of them can have real business potential as services, while others are likely to be of interest to hundreds of other communities. As the 3G network operator aligns with the communities, it also helps set itself as the preferred mobile operator for communities overall. The 3G network operator should even try to extend any kind of group discounts such as family calling plans and business VPNs to any community which cares to identify itself. While typically not the primary criteria, a community-positive operator can well be a reason to select between two otherwise similar network operators, helping capture customers and retain existing ones.

8.8 Mobile chat

Still another way we can extend ourselves into our community is chat. Chat in the 3G context is not 'talking' but very similar to 'chat' services on the fixed Internet, as a typed simultaneous communication usually with a number of participants. The fixed Internet chat board is a continuously updated listing of what everybody has recently said, and the latest addition appears as another entry. For those who have never been exposed to Internet chat, here is a small imaginary example of three friends chatting. Note that the first message is on the bottom, the newest one is on the top.

Communication in chat
Nicole (at 09:44)Yes, it was excellent! You HAVE to see it.
Nicole (at 09:42) Thanks. It sounds like a good place.
Janne (at 09:41) Nicole, did you see the movie last night?
Russell (at 09:38) I'm here. How about the new club 'Theatre'
Nicole (at 09:36) I dunno, lets ask Russell, Russell are you around?
Janne (at 09:32) So where's the partying action tonite?

Several factors make chatting distinct from any other communication, such as e-mail, SMS text messaging, teleconferencing, etc. As you can see from the above simple example, there are multiple 'threads' of discussions interviewed in a typical chat session. In the above example there are two threads, the discussion about where to go tonight, participated by Janne, Nicole and Russell, and the other discussion about the movie last night, by Janne and Nicole.

In fixed Internet chat – and mobile chat – people may join in and leave the discussion in the middle. The comments are usually always very short, 'chatting' rather than deep discussions, hence the name chat. In chat it is quite common to communicate anonymously or with a nickname, or even with an assumed identity. Chat is also very strongly an age-related communication means. The currently chatting generation tends to be the under 20 year olds, but they will bring the method to their working lives and eventually also convince many of their older colleagues and even bosses to use chat in some instances.

8.8.1 Chat and community

The community-building nature of chat is very powerful. If you were not there in the beginning, you miss out on what was said, and you might catch the discussion in the middle. Looking at the above example, if you joined the chat at 09:40, you would have missed the bottom three comments. Thus you would not know who Nicole was thanking and what place is being discussed. And until 'Janne' makes another statement, in some systems you do not even know that 'Janne' is live in the system, you might think it is only Russell, Nicole and you.

Because of the 'I may have missed something' aspect of chat, it is a very powerful means to build communities. But also the connection to the community via chat is very light, and can be severed easily. So for strong communities there should be many other means to communicate and ideally to meet with the community to strengthen the connection to it.

In the basic chat service there is no additional content provided by the service provider or the 3G network operator. The chat feature is enabled and those subscribers who log on will be charged for chat use, usually by the minute on line. The revenues generated are a direct function of how many chatters are attracted to the service. As chatting is a new way of communicating, and has addictive qualities, the 3G network operator should invest in marketing mobile chat and teach users to make use of it. Chatting should be promoted to early adopter types who are involved in many communities. Soon those who have tried chat in one group, will start to bring it to other communities as well.

8.8.2 Public bulletin board

Mobile chat can be engaged in today on chat-enabled mobile phones on networks where it is activated as a service. A variant of mobile chat is also the use of various public bulletin boards where chat messages can be seen, such as a large screen display at a nightclub, etc. There clubhoppers can send comments about the music or the nightclub can play interactive matchmaking games, etc.

8.9 Adult entertainment

One cannot complete a look into entertainment for a new media, without mentioning the biggest and most profitable parts of entertainment in *any of the other media*. The adult entertainment industry is a very profitable part of most media including magazines, movies, video rentals, the fixed Internet, etc. Adult entertainment has been the first to make money on practically all media, including early film (and peep) shows, printed pictures and postcards, pictures on playing cards, live theatre and burlesque shows, comic books, video tape rentals, etc. etc. etc. The adult entertainment services are probably the *most lucrative* part of the fixed Internet. Definitely early on the adult entertainment part of the Internet was the only one to make money out of the fixed Internet. The industry is quite secretive and precise data are not readily available. Yet the adult entertainment industry has pioneered many delivery and billing innovations on any technologies, for example in the Internet they were among the first to build in automatic links to forwarded sites, free but sensored previews of selected content, various payment methods, age screening, etc.

Video phones will become more and more popular with UMTS offering a range of services. The miniaturisation of camera technology is driving the market towards digital imagery that is part and parcel of the mobile phone package. In the not too distant future all UMTS phones can be expected to include a digital camera for either still or moving pictures. This example is of an existing NTT DoCoMo 3G FOMA video phone from NEC.

It should be noted that the adult entertainment industry is one of very few industries which has been able to generate value-add services *to fixed telephony*. For most other services we are not willing to pay significant premiums per minute to access for example the directory listings, or train timetables, or other content on a fixed phone service. But the business in adult services on fixed telephone premium services is so profitable that in most markets where they are legal, adult voice services (phone sex lines and adult chat lines) advertise their phone numbers on local TV at night.

Hint to all in 3G: learn from adult entertainment. The adult entertainment (sex) industry is the fastest at making money out of new opportunities. Make sure you keep your eye on the adult entertainment industry and learn from them, adopt their methods when appropriate.

The adult entertainment industry is looking to the opportunities of the high resolution colour video screens and immediate connectivity aspects of the 3G environment to deploy their services there. The adult entertainment industry is likely to be one of the early profit areas of 3G, of course with local regulations, and to the degree the industry is active in a given country.

When thinking of the potential for adult-oriented entertainment on the 3G platform, the 3G phone and network offer an almost perfect delivery system for adult entertainment. The service can deliver colour pictures, sounds, and moving images, offering the best of the realism of TV and video tape. Each use can be separately billed offering the best of billing of phone services. The service can be delivered around the world to a worldwide potential audience, offering the best reach of the fixed Internet. The users will have a portable (pocketable) device which can access the service from anywhere and can be taken with you into hotel rooms, etc., giving much of the portability benefits of magazines.

In many ways the 3G service is actually better than the best mechanisms today, also for the user. The user does not have to go looking for a magazine stand to buy *Playboy* or similar magazines, the content can be immediately accessed from the hotel room or wherever one might be. The user does not have to subscribe to the hotel's adult films offering which may be embarrassing on the hotel bill. The user does not have to take expensive remote PC access to the Internet such as the expensive Internet connections at hotels, etc.

For the 3G operator, adult entertainment will be a delicate issue. In some countries pornography is strictly illegal, and in many countries its offering on a public network could be seen as morally wrong. The 3G operator might want to keep a distance from the adult entertainment offerings, and only provide access to any such sites, and have the adult entertainment industry set up its own independent sites, portals, etc. The operator might consider setting up separate companies to deal with these types of service providers, to distance its brand and image. These could well have credit card based billing systems to further distance the 3G operator from the billing. Of course in some countries with liberal sexual laws, such as Germany, the Netherlands and Denmark, it might be seen as perfectly acceptable to have the adult entertainment site usage fees showing up on the regular phone bill and the operator could tap into significant early 3G revenues.

8.10 Real services today on Me

The services described in this chapter are not far-fetched fantasies, but actually very close to reality, and some may appear before this book makes it from manuscript to printed book. Many of the ideas discussed in this chapter do exist in more primitive forms, on services based on SMS, WAP (Wireless Application Protocol), and as features of mobile handsets.

8.10.1 Share digital image

Some Nokia Communicator users have been using separate digital cameras to take digital images and then move them to the Communicator via infrared port, and then send the pictures further to other people with Communicators – or to sites on the Internet, for example as attachments to e-mail. So the ideas of sharing images, taking pictures and sending them further via wireless means, is already happening, but in a very small way.

8.10.2 Image messaging

J-Phone in Japan has introduced the ability to use a built-in digital camera and messaging to send images directly from one mobile phone to another.

8.10.3 Mobile check-in at airport

Finnair has introduced the ability for its frequent fliers to check in via mobile phone and select seats, etc., and passengers are able to do the check-in as they approach the airport for example while sitting in a taxi. The service is seen as serving the loyal customers and saving them time. At the same time the airline saves in check-in handling costs.

8.10.4 Mobile chat

Mobile chat has been possible on SMS for several years, and numerous mobile chat services exist around the world.

8.10.5 Video conferencing

Video conferencing exists on fixed networks but mostly involves costly special equipment and often setting up two ISDN connections, etc., resulting in cumbersome deployment. Especially after the terrorist attacks of 11 September 2001, video conferencing has been growing fast and replacing some business travel.

8.10.6 Composing music

The ability to compose professional quality music will need the MIDI interface. But very simple music composing has been possible on many mobile phones for many years already. And at least some Japanese manufacturers have said they will be introducing MIDI standard mobile phones very soon.

8.10.7 Chat on a public bulletin board

In Finland broadcast television stations have started to run all-night chat bulletin boards based on SMS. The bulletin board is active after regular TV broadcasts end for the evening. Young people are providing all-night chat long chat messages ranging from 'is anybody awake' to 'I love you' to messages supporting local sports teams, etc.

8.10.8 Personals ads on mobile

Numerous mobile dating services have emerged around the world, providing a similar utility as personals advertisements in newspapers and dating services on the fixed Internet. In other words 'Single Male seeks Single Woman who likes to ...'

8.10.9 Attraction server

A novel variation of the dating services is the anonymous attraction server in Finland, being copied in several countries. The service allows sending an anonymous 'someone is attracted to you' message to a recipient, who is asked to test if he/she is attracted to the same person who sent the message. If they match, both are notified.

8.10.10 Adult entertainment

Of course adult entertainment exists for mobile phones and some of the most successful seem to be in Germany. SMS- and WAP-based services tend to be text 'stories' but the service providers are eagerly awaiting the availability of GPRS (General Packet Radio System) to introduce good resolution colour images.

8.11 I'm that type of guy

This chapter has examined the way services can build revenues and profits from the Me attribute of the 5 M's of 3G services. It is important to allow services with the Me attribute to enrich how we express ourselves. Also the Me attribute allows us to extend the self into the community, to build ever more attractive services for each individual in the 3G network. In building friendships, relationships and communities the 3G user will want to make the others feel good and happy. The 3G service provider can help in that.

The Me attribute is the most personal of the 5 M's. It is why we will view our services, our sounds, clips, news updates, etc., as so much more important than any other set. If well served, the Me attribute builds loyalty to our service provider. As most of the services in this chapter are person-to-person communication, there is not that much revenue to share as there is with more content-related services. The content here tends to be user-generated. But the 3G network operator plays a critical role in enabling new ways to use the mobile services to enrich ourselves and build our communities. Perhaps a guiding thought could come from Napoleon Bonaparte who said ''The only conquests that are permanent and leave no regrets are conquests over ourselves.''

9

The Profits of Money Services:

Expending financial resources

The ability to spend money and to use the mobile terminal for money transactions goes to the heart of the whole business opportunity in 3G. Some aspects of Money services, such as mobile commerce (m-commerce), mobile banking (m-banking) and mobile advertising (mAd) are such broad topics that a book could be written about them alone. For the most part, almost any service offered on 3G should be built to be billable, or else it should have the revenues arriving from some other source such as mobile advertising. This chapter looks not only at m-commerce and mAd, but also at other ways in which the Money attribute can be built into 3G services.

The services described in this chapter have a high benefit on the Money attribute. Most mobile services should have a Money component, and thus almost any new mobile service could be classified as a Money service. This chapter looks at areas where the Money attribute is particularly strong, and even there we have a chance only to explore a few of them. As such they tend to be services where the content is of considerable value and the mobile phone is only one, albeit perhaps the best, way to consume the content. Thus the operator is likely to receive only small proportions of these revenues.

This chapter will discuss some services which have a strong benefit from the Money aspect of a 3G service. The services are not in any order of importance and this brief discussion will not be able to adequately address even the major areas of Money type services. But a deeper discussion of a few services is useful to understand the nature of revenues and profits from the Money aspect of 3G services.

9.1 m-Commerce

Mobile commerce (m-commerce) is already happening in many small forms, from purchasing Coca-Cola from vending machines and paying for it with the mobile phone, to purchasing train tickets and even airline tickets via mobile phone. Still, m-commerce is in its infancy. There are several concurrent trends which support the growth in m-commerce. The current digitalisation and automation of the sales process is already expanding worldwide in the e-commerce and e-business trends relating to commerce on the fixed Internet. That commerce has been a viable business proposition only from the mid-1990s. This e-commerce-enabled trading community is much more easily ported to the m-commerce environment than the initial work needed to set up the e-commerce possibility.

The fast spread of Internet-enabled mobile phone handsets is creating an m-commerce-enabled community which is expected to exceed the total PC-based Internet community within a year. These users have WAP (Wireless Application Protocol), I-Mode and GPRS (General Packet Radio System) phones today, and will also have 3G phones starting in 2002. When most of the devices accessing the electronic commerce opportunities are mobile phones rather than PCs, at that time also the electronic commerce content and activities will be geared towards the buyers using mobile handsets.

While it is easy to associate m-commerce with consumer purchases – and thus being a B2C (Business-to-Consumer) opportunity, the B2B (Business-to-Business) side of m-commerce is as valid. Early on many of the users of advanced mobile phones will be business users and their needs will be among the first to be addressed by 3G services. The early B2B opportunity was analysed by Accenture in their 2001 report on the Future of Wireless where they showed which areas of the business purchasing activities are easily suited for m-commerce.

m-Commerce can easily be categorised as 'small transactions' and almost dismissed as a totally marginal part of the economy. There is no reason to assume that consumers would not be willing to purchase more costly items via mobile phone, just as they are doing on the fixed Internet today. Numerous early examples would include music CDs, books, electronics, and so forth. Let us start with something where the Movement attribute is also strong to support m-commerce, in ticketing.

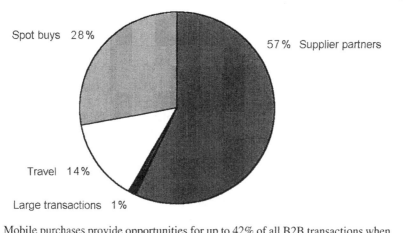

Spot buys 28%

57% Supplier partners

Travel 14%

Large transactions 1%

Mobile purchases provide opportunities for up to 42% of all B2B transactions when counting travel and spot purchases
Source: Accenture Future of Wireless, 2001

9.1.1 Ticketing

A very natural instance of monetary benefits of 3G services is any instance where currently people stand in line to buy tickets. Ticketing is a very promising early application for the mobile Internet as we often have to make an extra journey – or arrive early – or make advance booking calls – to get tickets (and thus waste time). In addition to that, we also often have to stand in line to pick up the ticket we have ordered or purchased.

3G services will radically change most instances where we currently stand in line for tickets. We can simply log on to the event's site or a general ticketing site, and select the time of the performance we want to attend, and select seats from those which are still available, and then 'click' to select and that is it. We don't have to call and wait in line for the next available sales person. We don't have to travel to the ticket counter to buy our tickets, nor to pick up tickets we have ordered. The ticket will be issued automatically to our phone, and we simply show up at the right time with our mobile phone. The ticket will be there, ready to be validated.

This type of mobile ticketing will be commonplace for movies, concerts, theatre, opera, the ballet; but also sporting events, conferences, transportation such as trains, planes and ships, etc. For receiving the service (entertainment, event or trip) we don't need to have a physical ticket printed on paper or cardboard; all we need is the permission to consume the entertainment or travel. That is what tickets have been used to prove. Now we can do that on the mobile phone, making physical paper tickets increasingly obsolete.

Who gains in mobile ticketing. A major cost of the administration of the money side of any entertainment or travel is the physical person who handles the money and sells the ticket. The handling costs of selling a ticket can be enormous. Electronic means to handle the ticketing shift part of the work to the person who places the order via mobile phone (self-service) and of course the other part of the equation is automation. By interacting with a ticketing software package, the person ordering the ticket gets exactly what he ordered, can easily get better service than what a human ticketing agent, bored after a long day of repetitive work selling seats to a play, might be prepared to do. And the savings in the ticketing human costs easily offset the much lower cost of the time and traffic that the person ordering the ticket will place on the 3G network to access the automated ticketing system.

Hint to service developers: look for people waiting in lines. Any place where people are standing in line or waiting on electronic queues, are opportunities for 3G service creation and m-commerce.

The money would be divided between the 3G operator and the company which wants its tickets sold this way. Early on, as the first such systems come on line, the 3G network operator could expect to take some portion of the money above the strict cost of transmitting the data and handling the billing from bringing the know-how and any marketing and branding involved. Eventually the portion of the ticketing revenue that the 3G operator might expect to get would approach that of being only a billing and transmission connection to the transaction.

Mobile advertising is likely to play a major part of the money side of ticketing. For almost any given type of ticketing need, there is some party which might be interested in advertising to that ticket buying person either before the purchase, or during the transaction for promoting another event or occasion. For movies it could be recently opened movies which might be promoted at the ticketing site, trying to steer undecided movie goers to buy that movie's tickets. It could be cross-selling, such as if we buy tickets on Air France to Paris, there might be an interested hotel chain in Paris to try to sell hotel nights, etc. Mobile advertising will be covered later in this chapter.

9.2 Selling digital content

As the 3G service environment is naturally digital, and it is a digital transmission medium, it is also a natural for delivering any digital content. With the 3G network's built-in ability to handle any kind of billing from the tens of thousands of dollars to fractions of a penny, the 3G network has competitive advantages over most other digital distribution systems. A significant issue for the content owners is digital rights management, and the various contracts, distribution agreements and participants in the distribution of content will need to adopt to the 3G converged digital world. The 3G environment provides a lot of opportunities but it also will face suspicion and even hostility as some will fear that their role will be removed in the new 3G environment.

9.2.1 Selling mass audience content which is already digital

The 3G network is the ideal vehicle to sell content which is created into digital form to begin with, is consumed digitally, and is primarily intended for mass-market sales. Thus many of the products that we consume via digital means such as music CDs, computer programs and video games may find that the 3G network is the most cost-effective delivery mechanism for their content. Currently most of those are delivered via various types of compact discs (CDs). The cost of copying CDs and boxing them and delivering the CDs to stores for resale add large portions to the cost of the digital content. The 3G network can cut the distribution costs into a small fraction of the current methodology. There is considerable opportunity for savings via the disintermediation of several unnecessary middle men. Looking at the distribution cost of digital content, in most cases the content could be delivered in a low QoS (Quality of Service) class in 3G, and thus not put as high a load on to the network as for example sending streaming music or live video feeds.

9.2.2 Selling know-how in digital form

While the 3G network is ideally suited for digital content intended for the masses, it can almost as easily accommodate the sales of digital content for more custom digital content. Many companies involved in selling information, knowledge or know-how, have come up with various systems and models to 'productise' and sell their know-how. The simple example is the law practice. With the advent of printing, it became possible to print sample contracts into booklets. Then when the PC became widespread, some lawyers put 'boilerplate' legal forms and contracts on to diskettes for sale to individuals at a tiny fraction of the cost of hiring an attorney to draft up the simple contract or will, etc.

In a similar way many specialists who sell mostly information and know-how, such as consultants, attorneys, doctors, engineers, architects, investment bankers, etc., are offering some of their services in digital forms. The problem still today is that there is no ideal delivery vehicle. The computer software stores and bookstores cannot carry every potential small-market specialist solution, in fact many computer software stores are increasingly becoming computer game stores. On the fixed Internet there is the added issue of payment and delivery. But as we have seen, the 3G network can handle digital delivery and the payment. And there would be no problem whatsoever to offer digital services for small niche markets. It is the best of all worlds. There will likely be a vast universe of specialist information and knowledge offerers, where the cost of the information will depend partly on the quality of the information, partly on time relevance, and partly on the brand power of the information producer.

9.2.3 Delivering digitally non-digital content

While digital content is a natural for the 3G networks, much non-digital content can be delivered digitally. Or in many cases the content is consumed in a non-digital way, but actually is produced at some point using digital tools. A good example is the modern newspaper. In most major newspapers the layout and printing is done with digital devices, while the newspaper is still mostly delivered as paper. Of course many newspapers also publish an Internet version, and some even provide content on mobile networks. Since the content has been digital at some point, the costs to convert to digital delivery are totally marginal compared with the overall costs. The beauty of digital conversion systems is, of course, that the majority of the conversion formatting work needs to be done once and then it can be repeated countless times to convert subsequent editions to the same format.

The same is true in many cases of TV broadcasting, in some cases also the motion picture industry, and many others. The content is increasingly processed digitally at some stage. The kind of content that could be covered is anything that could be consumed via a digital screen such as movies and TV and video tapes, magazines, newspapers, and books, etc.; or via digital speakers such as radio programming, musical concerts, training, presentations, etc. Any content on any of these media could probably be converted to 3G delivery. It is not necessarily the best medium to view a 3 h movie but imagine if you are stuck at a crowded airport and your plane is having an emergency change of its engine taking 6 h. At that point even a 3 h movie on a small screen might be a much better way to pass the time than having nothing to do. For the most part the 3G terminals will be best suited for short clips, and it is very likely that most content for the small screen will be such clips.

9.2.4 Content which could as well be digital – government fees, licences, taxes

Another area is content which is not necessarily digital, but could just as well be. 3G will be the 'final straw' to bring some of these types of services into the digital age. This includes any and all government fees, licences, taxes and other payments and forms that need to be filled. Considerable savings can be achieved by having the citizens fill out electronic forms themselves. America has been leading in the area of government by Internet, and m-government is an even better way than that. Mobile Internet devices are going to be cheaper than PCs so access to digital government will be possible for ever larger parts of the population. m-Government will include easier handling of payments, especially small payments.

9.2.5 Tangible item purchases

Buying a book or an electronic gadget, etc., via an m-commerce merchant will not bring drastic new benefits beyond those already discussed in the micro-payments chapter (Chapter 4). Suffice it to say that in most cases of regular commerce, soon we will have also the m-commerce versions of those shops and stores, and be able to purchase most items also with our mobile phones. Here the value brought by the mobile service would be equivalent to the utilities of the fixed Internet and the credit card, and the mobile operator would not be able to draw much more than the percentage that credit card companies gain today.

9.3 Mobile advertising (mAd)

During the year 2000 most 3G network operator business cases started to show mobile advertising as a new revenue source. The early examples centred around the location information and applied some Internet banner advertising logic to the assumptions. Mobile operators and analysts had as much as 10% of an operator's revenues coming from this new and untried revenue stream. During this 'Stone Age' of mobile advertising several horror scenarios were circulated around the theme of mobile spam advertising. During 2001 the Wireless Advertising Association and various other bodies laid down guidelines on what constitutes acceptable advertising and promotion on mobile devices. This part of the book will examine how advertising has evolved from that 'Stone Age'.

9.3.1 The Stone Age: mAd spam, i.e. location-based unsolicited push promotions

One of the first examples of innovation was the idea that location information and the needs of local stores at malls would be a natural match. That people could be greeted when they walked into a mall with various 'push' ads on specials of blue jeans on sale at the Gap, etc. Similar ideas have been suggested to be used when potential customers are close to stores and shops, possibly giving local maps and information how to get to the store nearby.

> **Hint to all readers: understand the end of the Stone Age**. Make sure that you understand how mobile advertising has moved past unsolicited location-based spam.

It is quite understandable that there is a strong attraction towards this type of targeted advertising by merchants. There is likely some novelty value to this type of advertising, so early on some may even be welcome by random pedestrians and mall shoppers. Merchants also may feel that this is a way to increase the traffic to their stores. But much like the reaction to junk mail at home, and spam e-mails at work, very soon unsolicited generic push ads will result in rejection by the consumers, even reaction against the cellular operator, the advertising store, and the shopping centre, etc. If the ads are location-based spam, it may even result in certain consumers actively *avoiding* given locations and routes simply to avoid the unwanted ads.

This is not to say that location-based ads cannot be successful. It is only that the 'walking down the street and getting a sudden ad from the neighbourhood sporting goods store' variety mobile ad will not be the predominant type. Location-based ads can be very big business as we see in the next example.

9.3.2 Location-based advertising at mass audience events

One of the best opportunities for user acceptance of location-based advertising stems from mass audience events. Imagine Madonna performing at Wembley Stadium in London. As fans walk in, they get a personalised message like this:

Dear Mr Ahonen, thank U 4 attending my concert. Here is a coupon for 20% off my new CD. Download it directly or forward the ad. Enjoy the concert. Madonna

---

The beauty of the mobile ad starts from the mass event and location-based services. Even with current technology it is possible to isolate the audience to be approximately those at a concert venue. With 3G it will be possible to make the targeting so precise that people inside the stadium will get the ad, but people outside at a McDonald's restaurant, would not get the ad. Thus it will be easy to target the mobile ads only at people who are actually inside the stadium. And for a Madonna concert, those who are inside are die-hard fans who bought expensive tickets to see the singer. They will love to get direct messages from the star.

What is in it for the fan? The fan gets the personalised message which many may save and cherish all by itself. And the fan gets the discount coupon, and those with advanced phones, get to download the songs directly to the phone.

What is in it for the operator? Traffic. About 15 songs of digital music to that portion of the audience who decide to download can be a wonderful windfall of extra traffic, centred in one predicable location. Ideally the transmission can be simulcast, allowing considerable further savings in distribution. And the operator gets traffic of mobile ads directly to about 15 000 fans, again simulcast. There is likely to be a large amount of follow-on traffic directly related to this ad campaign, in forwarding the ads and just showing the message to friends.

What is in it for the record label? They get the chance to digitally distribute the digital songs directly to the user, without printing CDs, packaging them in cases, shipping them to warehouses and stores, and paying stores the commissions. With 3G the record label is very sure of who was the final consumer, and the label will know it gets to be paid.

Most of all, what is in it for the artist? In this very classic win-win-win-win situation, actually the artist – in this case Madonna – gets the most. Madonna gets manageable direct contact for about 15 000 of her best fans, per concert. What is the value of these mobile phone numbers, especially over time? They are the ultimate brand-powered direct marketing channel to passionate fanatic consumers. The power and utility of this database is beyond anything that any artist has ever had before. Just imagine.

The next time Madonna visits London, she can send all these fans a message giving them first choice of the same seats as before – or even better seats this time, and a small advance booking discount. With a little viral marketing, Madonna might be able to sell out a whole concert with these 'leads'. Direct sales of subsequent concert tours, at a trivial cost of promotion. But it would not stop at concert tickets. Madonna could release her following CDs electronically first to her concert fans. It would help build a virtuous cycle – those who attend her concerts get music first – and those who buy music first get early warning of upcoming concerts. But it would not stop there. Madonna could use these to promote her music at a tiny fraction of her marketing costs. She could release her newest hits first to these registered fans for example as ring tones. These fans would be encouraged to share and forward the songs, creating a large alternate marketing channel for promoting her new music. The cost of promotion would be a tiny fraction of the millions that are spent today to promote any new song of any megastar.

But again, it would not end there. Madonna could expand her brand through this channel. Almost anything she would want to release as a 'Madonna product' such as T-shirts, books, fragrances, etc., could be marketed to these fans, as long as she kept giving some exclusive benefits, and did not over-expose herself via this channel. Even all of this only scratches the surface. The key is that those first 15 000 fans at Wembley were among the most passionate, loyal, and wealthy among Madonna's fans in the UK. Replicating this pattern across France, Spain, Germany, Italy and dozens of other countries, would yield a community of committed and affluent fans who have shown a willingness to pay much for Madonna experiences. Her next song could become a number 1 hit simply by activating this core loyal audience in a targeted campaign.

9.3.3 Not only Madonna or rock artists

The above example was built around a music artist. Let's ignore music. The exact same principles apply to any mass audience event. For example any kind of spectator sports, the bigger the better. Sports like football, basketball, baseball, hockey, tennis, golf, motor sports such as Formula One and Indy Car racing, track and field events, winter sports, water sports, etc. etc. etc. Most sports fans are fanatic and passionate about their events, teams and stars. Fans of Manchester United, or Ferrari or the NY Yankees or Montreal Canadians or Boston Celtics, etc. are true to their teams and will show their loyalty through colours, clothing, even affinity credit cards. These would love to receive direct messages from their sports club or its stars.

And how about all the fans who are not receiving the ads. If the mobile advertising is targeted well enough, it will appeal to all of the fans of that sport or activity. It will create a feeling of 'I want to be connected, too!' with all who are outside the distribution. As the fan who gets the ads shows them to his friends, they too will want to receive such targeted ads about their favourite activity. Here again we see community strength, and easily will also see viral marketing.

But beyond sports, there are mass audience events of passionate participants, such as conventions for various industries, religious gatherings, movies, theatre, concerts, ballet, musicals, opera; circus and magic; conferences on industries such as medicine, telecommunications, accounting. Masses get together even for things like outdoors such as the Boy Scouts. Many industries have mass events such as the auto shows, fashion shows, etc. Almost any significant hobby or business will have some kinds of mass gatherings. And these events could be exploited and location-based ads delivered for their participants, members and fans. Mobile services could be built to cater to these events and people attending them. In most cases probably at least some of the content could be converted to digital to provide new benefits to the participants. Of course at least tickets to the events, feedback forms, and future event brochures could be provided via the mobile phone.

9.3.4 Ads in music streaming

A natural part of advertising which has considerable potential, is that of ads amidst streaming music. Of course it will be possible to offer music streaming as a subscription service, but as it eats a relatively high amount of resources, and music streaming can easily run long periods, half an hour to several hours, this is not a very profitable proposition to the 3G network operator. Or else, it might be prohibitively expensive to the individual user. But radio-style advertising is a key to music streaming. In most countries where advertising on the radio is commonplace, that same model can be transferred to the mobile music streaming service.

Just like on radio, there would be a series of songs, then advertisements, then further songs. There could be disc jockeys (DJs) but not necessarily, as the mobile music streaming could easily be built to deliver the name of the artist and song to be shown and the DJ would not necessarily be needed. Music streaming has technical limits, as it eats up a lot of the bandwidth and thus is not as inherently lucrative as many other services. Music streaming has also several competing high quality delivery solutions, most obviously FM HiFi stereo radio. Many early mobile phones are likely to incorporate FM radio into the phone handset providing a viable competitive audio resource. But music streaming on IP (Internet Protocol)-based worldwide music sources could provide thousands of 'radio channels' of very specific and targeted music styles, especially for those whose tastes are not mainstream, and for all who travel and do not have access to their favourite music when in other countries.

The money equation in music streaming is interesting as it has a lot of potential to salvage royalty revenues from an industry scared by piracy and the advent of peer-to-peer music sharing pioneered initially to great success by Napster. With 3G music streaming there is of course the issue of digital rights

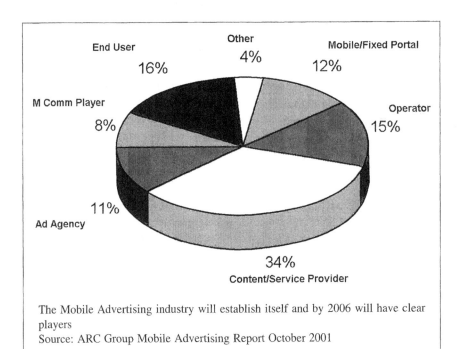

The Mobile Advertising industry will establish itself and by 2006 will have clear players
Source: ARC Group Mobile Advertising Report October 2001

management, the owner of the song and artists involved will want to be paid of course, else they will not release their music. But the use of the 3G network to deliver music will also demand a large payment to the network operator.

9.3.5 Your absolutely favourite music

The most interesting potential of streaming music is the targeting. As long as both the song digital rights owners, and the delivery network load are fairly compensated, there is no reason why not all music *in the world* could not be accessed. And simple digital DJ software algorithms and music classifications could very easily produce very targeted and personalised music streaming channels. For example there could be channels which differentiate between old school and new school rap music. Or hard rock as distinct from heavy metal, etc. That kind of targeting is not viable in most radio markets, where in a city of half a million people, perhaps one black music/dance music station could cover rap partially, and one or two radio stations could cover rock music in general, but not be able to focus on only heavy metal or only hard rock. And so forth. With 3G phone users growing to hundreds of millions, the accessible market is enormous, allowing for very tightly targeted music offerings.

The key is that once a channel is selected, there would be payments to the network operator and the artists, and the payment would be from the advertisers. The advertiser could benefit greatly by being able to separate the generic listening demographics and tastes from the actual personal data of the consumer. Thus even if a 42-year-old *white* man happens to like new school rap music – like I happen to do – the advertisements to me could still be targeted to my profile – correctly classifying me as a white 42-year-old professional telecoms consultant – rather than feeding me 'mass audience ads' for the under 20-year-old rap music listener, probably ads of Nike running shoes and Levi's jeans, etc. This kind of targeting is simply not possible in radio today. Note that I could be fed different ads in my ad break than a 25-year-old friend who is listening to the same channel.

9.3.6 Guaranteed listeners to ads

But the 3G network provides a lot of information on the listener which could greatly help in the viability of the business proposition. The 3G network could 'guarantee' that each streaming listener will hear advertisements and not be able to skip them by tuning into another channel. The ad proportion could be for example three 30 s ads after every five songs. On radio today, if you listen to music on your favourite station and you notice it went into an ad break, it is easy to hit a speed dial button and switch to another of your favourite channels, which quite likely is playing music again. In this way you can 'carousel' around channels and skip a lot of ads running on the regular radio.

On a 3G network, the operator could easily track how many ads any given subscriber has listened to (or viewed). And if you tune out at the start of ads, you could not listen to more streaming music – of ANY streaming channel – until you have listened to (or viewed) your required set of three ads. This is a great improvement for the advertiser over the current electronic media, where at all other media you can skip over, switch channels, or avoid viewing the ads. As long as the streaming music channel explains well how it works, and as long as the ad breaks are not painfully long, the concept is likely to work very well.

For the listener this can be very much better than any current radio channel. The streaming channels are likely to become very precisely targeted and categorised and classified. So rather than a 'rock music station' one could be a station focusing on mostly British heavy rock of the 1970s; another could be playing only the Rolling Stones while another could focus on music which *sounds* like British heavy rock of the 1970s but not played by British bands, etc. Now you can select and tune into your favourite music by your preferences and moods. The system could easily learn based on your preferences, and suggest channels based on your established preferences. 'Since you like the Beatles from their later years, you might want to try listening to the Electric Light Orchestra', etc.

When financed by ads to guaranteed listeners, the music could always be free, just like on radio today. But now you could have access to your favourite radio stations wherever you are. And regular radio advertising would find itself fighting against a newcomer which can deliver guaranteed audiences to the ad breaks, and also precisely targeted audiences to ads.

9.3.7 Rewarding advertisement recipients/frequent viewer miles

Of course not all people go to mass events or want to listen to music. Wireless advertising will be reaching us many other ways, and one of the most exciting aspects is rewarding the viewer of the ad.

From the production point of view, the cost of a direct contact with a viewer of an ad, via mobile advertising, is a miniscule fraction of the cost of current means, such as TV advertising, newspapers and print media, and direct mail. The savings in the cost of delivery of the ad will partially need to be used to ensure that the advertisement content is very appealing and entertaining or informing to the target audience. But another opportunity also arises, which has already been tested in Japan. It is rewarding the advertisement viewer by bonus points, such as frequent flier miles or other bonus plans.

The mobile advertisement would work so, that the mobile phone subscriber would initially indicate preferences of what areas may be of interest, such as travel or fashions or sports, etc. Then the advertisers would not send ads directly to the consumer, but rather a series of click-to-ad links. The links would be the short headers or 'titles' of the ads, such as 'British Airways weekly specials' or 'McDonald's coupons', etc. The links themselves would not cost anything for the user of the phone, and not cause any delays in the service. When the user would click on a link and view an ad, the user would receive a set small amount of points, such as for example 5 frequent flier miles. Of course viewing the ad would still cost the user nothing, except for time and inconvenience. By clicking on the ad many times, the same phone owner could not get more miles, but when clicking on to other ads, the user would.

Hint to 3G network operator: get your system ready to track forwarded ads. Make sure your system can report on who forwarded what, how many times an ad was forwarded, etc. That will be a big competitive advantage for you.

The user would never be forced to view ads. The viewer would get titles of ads only by preferences and interests, and if the viewer would click on to an ad, the viewer would receive a small but tangible benefit. In many cases the advertiser would probably like to use viral advertising and hope that the person who clicked on the ad would also forward it to friends. By forwarding the user would not get more points, but if the recipient would then click on to

Entertainment terminals like the Nokia 5510 will become more popular in UMTS as we see the combination of services into a single device. More categories of terminals will be developed that address these new markets.

the ad, the forwarding person would get another set of the same amount of points, for example another 5 extra miles. There would probably not be a limit to how many people one could forward the ad, as long as the points would only be awarded if the recipient did actually click on the ad, and as long as no user could receive more points for watching the same ad twice.

The benefit could also be in the form of free minutes of mobile phone airtime, or free film clips, sounds, games, etc., or any other benefit of that kind. Ideally the user could select how to be rewarded.

The actual business case behind this kind of rewarding the viewer is actually very positive. The cost of delivering the ad digitally directly to the phone as a link, and also paying for the short transmission cost of the actual ad of those who clicked on the link, is much less than reaching a similar amount of targeted people with any other means. The cost of the bonus points, or frequent flier miles, etc., for those who actually view the ads would still keep the case as a very successful and cost-effective targeted ad campaign.

9.4 When ads become content

We all can remember instances when a given television ad has been particularly funny or entertaining. The challenge for creative talent in advertising is to try to develop ads which would be seen so desirable that the viewer would want to see the ads, and not need other content. In an ultimate advertising world, a whole media channel could be created with nothing but ads, but if the ads were seen as useful content, there would be viewers/readers to the ads.

9.4.1 Examples exist from the non-digital media

There actually is some precedent to such instances, although they might not seem like it at first glance. Over 100 years ago newspapers were full of small classified ads of many kinds. At some point some of those ads started to migrate to a section of the phone book, and soon were known as the Yellow Pages of the phone book, which turned into their own section of the classified ads for the phone company. Early on the same phone book had white pages of names, addresses and phone numbers, and separate Yellow Pages of ads. Eventually in larger cities the Yellow Pages book became its own separate bound book. Some businesses tend to use only the Yellow Pages part of the phone book series, and of course some independent Yellow Pages publishers exist. This is one case where content is created completely by advertising. In fact there is separate advertising (colour pages, etc.) sold to Yellow Pages books, which themselves contain only advertising.

Another such development is the automobile and housing "for sale" magazines in many cities. Again, the classified ads were introduced into a separate newspaper or periodical, with no other content except the classified ads, and people pick these up for free, or even are willing to pay for such magazines in some cases. Again there is only advertising, no other content.

Yet another example exists, this from the area of popular culture and the youth segment. In the 1970s various rock and pop bands started to issue 'film clips' to help promote new albums and tours. Artists such as Queen, David Bowie, Rolling Stones and Abba were pioneers in this area. These were of course the early versions of what is now known as the music video. But for the artists and the record labels, they were an efficient advertising means to get the faces and sounds of the pop artists directly to their target audience, via television. These music film clips were purely advertising of the bands and their songs. Then, in the early 1980s Music TV (MTV) appeared. This was a channel with 24 h of music videos. It turned what were ads, into content. And of course we all know how powerful MTV has become with numerous national editions of the service around the world, and many 24 h channels broadcasting concurrently. And yes, all kinds of youth-oriented advertisers buy advertising time, to place their real ads, amidst the music videos.

So yes, it can be done, that content would be made up entirely by advertising, and in an ideal situation, it can be one of the most powerful and addictive aspects of the culture. Just to see how powerfully addictive it can be, try to permanently remove the Yellow Pages from your department secretary, or tell your 12-year-old child that MTV is to be disconnected.

9.4.2 Nightlife guide to the city

A service of free content consisting totally of ads also has been pioneered already for the mobile environment. In Helsinki, Finland, in the late spring of 2001 the local youth magazine *City Lehti* with a strong mobile and Internet presence, decided to launch the nightlife guide to Helsinki, targeted at the bar-hopping, disco-chomping nightlife. The users have to sign up for the service – for free – and then they receive several SMS (Short Message Service) or WAP ads over the weekend. These are of course ads from the major bars, pubs and discos of Helsinki; advertising their specials, hours, visiting DJs and bands, special theme nights, etc. The service was drawing a lot of initial interest from the club-hopping nightcrawlers of Helsinki.

The free content is a remarkably powerful tool as it has very strong winners in a 'win-win-win' situation. The end user gets valuable information or entertainment, for free. The end user is very likely to spread the word – which promotes viral marketing. The advertiser gets targeted audience contact, where the audience *wants* more ads. The mobile operator gets paid for traffic first from the sending of the ad, and then more traffic from the onward transmission of the forwarded ads. And the mobile operator gets associated with very much goodwill in delivering free content. Of all the potential for making money in 3G, I think this is the most challenging, but also in the long run the most beneficial for the whole industry. The industry should recognise and promote the best examples of this kind of free content. In fact I suggest that the 3G industry set up an annual award for best free content.

9.5 Viral marketing

As we saw in the Me chapter (Chapter 8), the community nature of mobile communications is one of the most powerful features distinguishing it from other media. We think of our mobile phone as an extension of ourselves and we contact our informal communities through this new means. Viral market-ing – where the recipient of an ad decides to send it onward to his friends – is one of the biggest potentials of 3G. The prudent 3G operator will be building the message delivery tracking of its network so that viral marketing activities can be tracked and accurately measured and reported.

Why is viral marketing so important. Because the first forwarding person is both interested in the subject of the advertisement, AND the forwarding person thinks that the target person is ALSO interested. Let's take an exam-ple. Let's say that Jane likes Swatch watches, and has a few. And she has signed up on a Swatch database (I do not know if one exists, so this is a purely hypothetical example). So Swatch might send her an ad about the new Swatch watches for the spring collection. Today it might be a printed brochure sent to her home, or an e-mail mailing sent to her e-mail address. In the future it could of course be a mobile ad.

If Jane really does like Swatch watches, she probably also knows several other acquaintances who also might be interested in Swatch watches. If Jane gets a paper catalogue sent to her home, she is probably not going to copy pages from it to give her friends. There is no viral marketing. With e-mail based ads and catalogues, there is some chance that she might actually re-send it further. But because of the vast variety of firewalls, differing e-mail systems, etc., it would be very difficult for Swatch to make an automated forwarding link for Jane, and to track where the ad had gone.

9.5.1 Click-to-forward

In 3G there is the natural ability to forward messages and pictures, it will be part of multimedia messaging, and examples were given in the Me chapter (Chapter 8). The 3G network operator should make it very easy to have a click-to-forward button on any mobile ad.

Even better, the system should tell Swatch (and the ad agency involved in producing the ad) how many times their ad was forwarded. The ad could even be designed to give some benefit to Jane for forwarding it, such as the points mentioned above. But the key is that Jane is likely to forward the Swatch ad *only* to people she thinks really want to see the ad. For Swatch this is a remarkable benefit. They only had Jane's name in their database. Even if there is overlap, in that many of Jane's forward target people are actually also registered, it is quite likely that Jane's knowledge of her friends is much more accurate than the accuracy of most profile databases for advertisers. Thus Swatch gets in contact with people who are true prospect buyers and who are not in their database.

9.5.2 Virally forwarded links

A variation of viral marketing is a sponsored forwarded link. If you liked the ad or sponsored content, and would like to send it to a friend, there could be the opportunity to forward a link. With the forwarded link there should be the chance to send a greeting to the friend. This could be in the form of the abbreviated SMS text message, i.e. about 120 characters out of 160 would be given to the sender for free, and the link and short message from the advertiser would be appended to the message.

The benefit here is that the advertiser gets a direct contact from the target person. If Jane sends the Swatch ad to Mike as a link, and Mike clicks on the link to receive the ad, then Swatch gets Mike's contact – and thus also directly his mobile phone information. Jane could be rewarded only if any forwarded links turn into contacts. And of course the 3G network operator would keep track of Jane's forwarding, making sure that she does not abuse the system by sending the same ad to the same person using that as a 'free SMS' messaging platform. The system could be built so that any one ad from any one person can only be sent once to any target person for free.

9 Vignettes from a 3G Future: Let's Pick the Movie Together

When my girlfriend and I are thinking about seeing a movie, we now use the free preview ability on the 3G phone. I call her, and then we both go to our movie theatre pages and click on a movie trailer. We can see about a minute's worth of an ad from the movie and it makes it a lot easier to decide which movie we want to see. It is then very easy to book seats, as the click-to-book button is right there with the service, and best of all, we don't have to stand in line to pick up tickets, as the tickets are already on our 3G phones.

The mobile purchase of movie tickets is likely to be one of the early popular services on 3G. Movies appeal to the youth segment. Movie producers and distributors will be eager to place their trailers – 1 min advertisements of their movies – on to the 3G screens. They are likely willing to provide the connection for free just to promote the movies. The movie houses are eager to automate ticket sales and also minimise needs to man ticket counters. The savings from these personnel cuts can be passed on to covering the airtime fees of handling movie listing requests and movie ticket purchases. The savings in ticket processing costs will probably totally cover the costs of the airtime and movie houses and 3G operators can make a very profitable set-up out of providing mobile movie tickets.

9.6 Call to action in mAd

A separate but very powerful aspect of mAd is the ability to have instant call to action. Various click-to-view, click-to-trial, and click-to-buy buttons and links can be built into the ad, generating immediate calls to action. These type of interactive abilities do not exist in most current forms of advertising. Combined with m-commerce the power of mAd becomes even greater. The target ad recipient can receive the ad, make a purchase decision, and pay for it all on one handheld personal device which is carried on the person all the time. The various issues around m-commerce were discussed earlier in this chapter, as well as in the previous chapter on micro-payments (Chapter 4), so they will not be repeated here. It is important, however, to remember that the immediate call to action is a particularly strong benefit to advertising on mobile phones.

9.7 User created content

Another major area of commerce which will be on 3G networks, is that which is wholly or in part generated by the user. This will include classified ads for almost anything we might want to sell, such as our apartments, cars, used household equipment, furniture, etc. etc. etc. User created mobile commerce content would also include auctions, even though both auctions and classified ads would require someone to set up the virtual entity to be the collection point for the service. This could easily be around a known brand, such as the local town newspaper for the classified ads, and a company like eBay for auctions.

9.7.1 Mobile classified ads

Mobile classified ads are likely to be an early service with mass appeal. The categories will mirror those in newspaper classified ads. The benefits for the media include the fact that the user gets to input the text of the ad directly, removing the service person from the costs. The benefits to the individual advertising include the vast reach of the millions on mobile networks, and immediacy. If I get the urge to sell my sofa, I can place the ad on a mobile classified ads service in a few minutes, and I might have the first call or text message within the next few minutes. Of course 3G networks and terminals will allow for convenient placement of electronic images of what is being sold. There would be a small charge to view the items, so they would probably be listed with a click-to-view link for 'image' and accessing that would cost the potential buyer roughly the cost of delivering the image. Certainly there would not be a payment to the owner of the item for viewing the image.

9.7.2 Mobile auctions

A similar use would be mobile auctions. With mobile auctions there will need to be administration of the site, guaranteeing quality and delivery of what is being bought, and payment to the seller, etc. But current Internet-based auction sites have found ways to resolve most of these issues.

9.7.3 Personal private banking

Mobile banking was discussed in the Moment chapter (Chapter 7). It includes the ability to bypass the banking establishment altogether, and send money directly to a friend's mobile phone account. This is of course another possibility in user-created mobile commerce services.

9.8 Real services today around Money

Money services are emerging all around the world, and it is becoming impossible to stay abreast of all the variations. A brief look into the situation in the winter of 2001–2002 should be sufficient to show Money-based services on the mobile phone are already real revenue generators literally around the world.

9.8.1 Coke machines and vending machines

One of the first curiosities of m-commerce were the experiments with vending machines first seen in Finland where users could purchase Coca-Cola and similar items by sending an SMS message to the vending machine and have the purchase showing up on the phone bill. Now these are commonplace in places so diverse as Hong Kong, Poland and the US.

9.8.2 Paying for underground tickets

Helsinki public transportation took into use the ability to pay for tram and underground train tickets by mobile phone from the start of 2002. The ticket costs the same as bought from a machine but the user does not have to stand in line at a ticket machine, saving time.

9.8.3 Movie tickets

Many forms of ticketing solutions exist in many countries. For example Colorado, US was one of the first places where it was possible to order movie tickets and pay for them by mobile phone. You need to show up at a special counter to pick up your tickets.

9.8.4 Tickets direct to your phone

The logical next step in automation is the elimination of the printed ticket. Austria was one of the first places to allow purchasing train tickets by WAP phone, and when the conductor asked for tickets, the traveller would simply show the phone. No printed ticket necessary.

9.8.5 Mobile discount coupons

Mobile coupons have been introduced in many countries, from Spain to Finland, Germany to US to Japan. Mobile coupons have been offering discounts on items from golf clubs to McDonald's to personal computers and peripherals.

9.8.6 More advanced music on mobiles

The ring tone business started in Finland in 1998 and has grown to over 1 billion dollars worldwide by the end of 2001. Now polyphonic (multiple simultaneous tones) ring tones, and integrated FM radios and MP3 players are appearing on newer handsets.

9.8.7 Advertising sponsored free news

Several countries have introduced free mobile news services sponsored by mobile advertising. Finland, UK and France are among the early ones to have free news such as headlines, financial, sports, etc. news delivered without any payment by the mobile subscriber, but at the cost of receiving typically one SMS advertisement per day.

9.8.8 Advertising bonus points

The idea of collecting frequent flier miles or other bonus points has been experimented with in Japan. In fact the Japanese mobile advertising industry is growing so fast, and their fixed Internet use is so small, that mobile advertising is expected to exceed fixed Internet advertising early in 2002.

9.8.9 Cell broadcast advertising

In Japan the first examples of using location and broadcast have been seen, as the cell sizes of the radio network closely correlate with the train station stops. The local daily commuter business is related to these train station stops, and therefore it is a good opportunity to advertise to all who are at a given cell. This could be specials at a local restaurant, or specials at the local grocery store, etc.

9.8.10 Advertising as content

In Finland a young adults-oriented free mobile service is providing ads for what is happening in Helsinki nightlife. The various nightclubs, bars and discos use it to promote their specials, bands, etc.

9.9 Got your money

This chapter has looked at some services which have a strong benefit on the Money aspect of the 5 M's. The services were not an exhaustive listing, and described mainly only to provide a deeper understanding of the types of 3G services that are expected to be created. Each of the described services would have several of the other 5 M's as a strong attribute as well. In designing 3G services the operators will need to experiment and be creative.

Potential services that can capitalise on the Money aspect number in the thousands – theoretically every service could have a Money dimension. And operators should not worry too much early on in trying to get it exactly right. With a new technology it is very difficult for users to give exact guidance on what they really want. But no service should be contemplated without ensuring that the Money dimension is covered. In business almost everything should, in the final analysis, be driven by the Money aspect, as the former CEO of General Motors, Frederic G. Donner said: ''When you come right down to it, almost any problem eventually becomes a financial problem.''

10

The Profits of Machine Services:

Empowering gadgets and devices

The last major area of beneficial services by using the 5 M's is allowing for machines to perform activities and communicate. This type of listing will probably be endless, but some early obvious areas are automobile telematics, home appliances, metering devices both fixed and mobile, robotics and voice-activated automation services. The revenue and profit streams relating to machine traffic are quite different from those that are generated directly by humans. Machines do not get sudden urges to play the lottery, surf the Playboy pages or fly to Paris. Machine traffic, on the other hand, can be controlled much more than human traffic, sending machine traffic at times of low congestion, etc., producing good revenue to network cost ratios, increasing the profitability of given traffic.

It is important to remember that most human accessed services can also be enhanced with the Machine attribute. This is one of the key ways in which 3G operators can increase their efficiency in delivering services. While that is also using the Machine attribute, and an important aspect of the money in 3G services, this chapter will focus primarily on services where the automated devices initiate the communication activity. The services described in this chapter have a high benefit on the Machines attribute.

There will be many opportunities to create automation and increase convenience for individuals, but also numerous opportunities to allow machine communications to increase business and government efficiency. Machines can communicate with other machines, for example, the car scheduling an oil change. Machines can take human interaction and respond, such as a translation service. Machines can also initiate contact with humans, such as voice broadcasting of political ads to phones. Any of these are covered in the broad range of Machine attribute services.

10.1 Automobile telematics

The first and most obvious application of machine-to-machine traffic and new services will be services around automobile telematics. The development in automobile telematics is not driven by the telecoms industry, but rather by the big automobile manufacturers, who have seen the profits in their industry dwindle over the decades. Each pursuing the target of manufacturing the world's most intelligent car, all major manufacturers are speeding up development of data processing and telecommunication ability of the car. The players are all the biggest manufacturers from Daimler-Chrysler to General Motors to Ford to Volkswagen, Fiat, Toyota, etc. Each of the luxury brands such as Jaguar, BMW, Mercedes Benz, Cadillac and Lincoln want to be known as the smartest car.

What can we expect. Navigation and safety features may come to mind to the driver, and service maintenance issues to the manufacturer. There are many solutions already being deployed to fleets of cars which include mapping software, intelligent driving instructions, and remote maintenance of problems, for example dispatching emergency assistance if the car airbag is deployed.

Of course mobile phones have been installed into cars as accessories for many years, so drivers can enjoy hands-free calling and use the car's built-in antenna, or the charging unit for the mobile phone's battery. Car manufacturers see a much greater potential for an intelligent car than just these few isolated examples. The car of the near future will be a very complex series of computers and communication devices. The car can of course do everything mentioned above and there are benefits from integrating those services. The actual telematics solutions will initially bring in hardware sales revenues and later the proportion of the service revenues will become an increasing part of the total solution.

10.1.1 Car service and maintenance

Probably the most mundane need seen by car drivers is the service and maintenance aspects of telematics. The car systems today have a lot of self-diagnostic abilities, as there are already several microprocessors and sensors throughout the electronics of the car. It is a small step from these to adding the communication ability and then having the car contact the authorised dealer and ask for a time for a check-up or oil change, etc. The system can be improved with integration into centralised calendaring and scheduling systems of the car owner and thus the system could look for a suitable moment in time, not only good for the garage, but also for the car owner's schedule. The telematics traffic load to the network would be very slight in such systems where the car would transmit a few bits of data to the centralised scheduling location. These types of telematics data traffic arrangements would typically be covered by the fleet of automobiles and negotiated as bulk contracts.

10.1.2 Car navigation

The next obvious application is navigation. Many CD-ROM-based mapping systems already exist, and some cars already incorporate GPS (Global Posi-

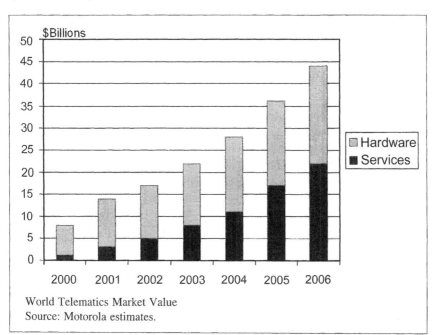

World Telematics Market Value
Source: Motorola estimates.

tioning Satellite) systems to bring positional accuracy to the mapping soft-ware. From here again, it is a small step to a 3G-based telematics system to bring more utility and assistance. The main problem with CD-ROM-based mapping systems is that they do not automatically update based on road conditions, repair works, traffic congestion, etc. A dynamic system which contains the real-time information on the road conditions and knows the exact location of the car and its destination, will be able to assist even more than merely a mapping system. Here the partners sharing in the reven-ues would typically be the car manufacturer and its built-in navigation system, and often a third-party navigation solution company. Either would negotiate a deal with a mobile operator to enable the dynamic updates.

10.1.3 Road advertisement requests

As the driver becomes hungry on a longer journey, he might ask the system for what fast food restaurants lie ahead, and the system could dynamically adjust to show only those restaurants over the next few exits. This would serve a similar need as the billboard ads at many motorways around the world, which say what fast food restaurants are at that exit. But the problem of course is that if you don't like those selections, you never know whether the next exit has any better selection – or in fact if the next exits have any restaurants at all.

For night-time drivers, such as the long-haul trucking industry, the open-ing and closing hours of the restaurants are also significant. You don't want to pull off the road when driving on a tight schedule, and find out that the fast food restaurant closed an hour ago. Similar services could be for service stations, hotels, shopping malls, etc. The main issue is that these should not be taking the drivers attention away from the road, so they should be not automatically pushed at the car system, but rather be available as pull services, where the driver (or passenger) requests information on what is of interest. The services should also be offered on a voice basis, through text-to-voice solutions. As in any advertising, the cost of delivering these ads would be borne by the businesses which want to advertise, i.e. the restaurants, hotels, etc. Here the business model would be a development of the billboard advertising companies and mobile advertising.

10.1.4 Car security

Car security is another area which can benefit greatly from 3G network services. In some countries where car theft is a significant problem, this could be a very high value extra benefit and soon deployed to most new car fleets, and quite possibly retrofitted to older cars. The car security can have features by which the owner's phone is contacted under certain situations such as the car moving without the owner (or his phone actually) being present, and the owner given the chance to remotely set the car to stop. This would need to be built to be intelligent, it could not simply apply the breaks while the car is in motion otherwise it could create a serious traffic hazard and another car could crash into the supposedly stolen car.

An intelligent theft-recovery system could first call the police and inform the police where the stolen car is currently driving. Then the system could wait until the police have caught up with the car, and then start to reduce the effect of the gas pedal. In this way no matter how hard you press the gas pedal, the car would just gradually slow down. Then the car thief could be told by the car that the police are behind the car, and that the car is going to stop. The system would not only stop the thief from getting away with the stolen car, it would make it very difficult to destroy the car as the police would be nearby, and it would allow the car to be recovered, and best of all, it would allow the police to catch the thief.

As the prices of simple IP (Internet Protocol) video cameras keep coming down, the car could even be built with a video camera with a view to the cabin of the car. This camera could take a picture of the thief operating the car, to act as proof and aid in identification. And in that case the police could call the car and talk to the known car thief, addressing him by name: ''Hi Joe, it's the police. We are in the unmarked car right behind you, and this car you have stolen will start to slow down – right now. Just pull over, there is no place for you to run, and we have your pictures of driving this car already stored on the network. You really should stop this car theft business, its getting to be too hard.''

With security systems a partial interested party is the car manufacturer who can market its new lines of cars being more theft-proof. The more likely major players in the security systems will be the various third party security solutions already available in all countries where car theft is a significant problem. These companies are likely to create much more advanced and sophisticated car security solutions than the simple ones described here. In high-risk countries a possible interested party could be the insurance industry which may want to support such developments and give discounts for car insurance in cases where the added methods are deployed.

10.1.5 Car multitasking

Telematics can also integrate the car further and help the driver be more efficient through multitasking. For example, the 3G system could connect to the e-mail of the driver and read out loud e-mail messages. With simple voice activation, the service could place read e-mails into categories which need to be replied to, forwarded, or deleted. Just like currently many drivers use the driving time to listen to voice mail messages, very soon people travelling in cars will be able to unload some of the perennial e-mail overload.

10.1.6 Games for kids in car

Another application for cars is gaming. Not for drivers, but mostly for children. The possibility to play Sony Playstation and Nintendo type games in the back seat, on screens situated at the back of the front seats, is perhaps the answer to the continuous children's questions of 'Are we there yet?' The games could be played independently, or against the other child in the car, or not far in the future, also against players elsewhere in the 3G network, for example a family friend who is on another highway also going to their vacation. But before that, probably cars will connect to game servers at hotspot areas such as at service stations while the car itself is being refuelled. A new game level might be downloaded at the same time for the next several hundred kilometres or miles of travel.

> **Hint to 3G network operator: if you are targeting car telematics, get a strong partner.** The special skills to build car systems and have them integrated take a major player or players. If you intend to be a major player in the automobile telematics area, get a strong partner in this.

10.1.7 Service integration

Eventually the integration of the personal digital device such as a PDA (Personal Digital Assistant), the mobile phone, the personal computer, the home entertainment system and the car, will bring best benefits to the intelligent car experience. The car system should know when the driver turns the ignition key, what has happened recently, and make intelligent assumptions based upon that. For example, if the family has been planning a driving weekend, and searching for some of the potential locations at home on digital TV, the car should be aware of these, and automatically suggest these, and routes to them.

10 Vignettes from a 3G Future: m-Translator

These real-time translator services just keep getting better. I remember some of the early systems only allowed a few seconds of speech at a time, and the translation was very clumsy. But now the system is so intelligent, it sometimes asks me which meaning do I intend, like 'here' as in this place or 'hear' to listen to, etc. It makes dealing with foreigners in our country so much easier but of course I use it mostly whenever I travel to other countries.

Translation systems are developing very fast and many simple voice-to-voice translation solutions exist. Soon major language groups will be covered and eventually all significant translation needs. The beauty of a translation system is that when built as a service on mobile voice, the network operator gets to keep the call within its network, connecting one mobile terminal with a fixed translation server. There is less load than connecting two mobile phones to each other, and of course there is no payment of interconnect to other networks. Yet the operator can bill the user at least the same cost as a regular mobile phone call, making such translation systems very profitable.

10.1.8 Personalised car services

The intelligent car could also quickly learn what are the main drivers' preferences, so that if the husband prefers to listen to stock quotes and a market
update, and be offered the route which has the least amount of traffic even if it
would take a little bit longer, the wife might prefer to have personal e-mail
read, and perhaps prefer the fastest route regardless of traffic congestion. The
car would set the appropriate air conditioning settings, play the appropriate
radio stations, etc. The car could even have a personalised speaking voice for
each driver according to their particular likes and dislikes.

10.1.9 Auto-pilot/built-in chauffeur

The car industry is preparing for the vision of cars being able to drive themselves. The first futuristic visions of totally automated cars were provided in
the 1950s and ever since the industry has been taking small steps in that
direction. Recently innovations such as parking radar and collision avoiding
radar have brought the car the ability to actually sense its immediate
surroundings and other objects in it. Some auto-pilot chauffeur solutions
have already been prototyped for proof of concept testing.

While the totally automatic car is still several years away, the car's utility
will benefit from 3G connections. As the car's processors will manage the
data such as destination, route, and en-route entertainment, the *road* may
want to communicate with the car. There may be instances of sudden alerts
and changes which the driver would not think of asking for – such as 'has
any tree been hit by lightning and fallen on my intended route causing a
detour and delays'. Such accidents and changes in the environment can be
transmitted directly to the auto-pilot of the car, which could then inform the
owner of the car and ask whether the car should compute an alternate route.
The interested party in such cases tends to be the authority governing the use
of the roadways, and likely first use will be in congested high-tech areas
such as in the state of California. The state could pass tax laws to promote
the adoption of intelligent devices in the cars, and the roads built with the
communication ability, to reduce overall road congestion.

10.1.10 Who makes money in car telematics

The data traffic that the car would transmit and receive for its diagnostics and
security and alarm instances would be very light in traffic load. The maps and
driving intelligence would be mostly stored in the car, not typically down-

loaded from the network. Some update information such as where is traffic congestion, or perhaps views of selected intersections, etc., might be transmitted, but these still would be a relatively small amount of traffic when compared with Internet surfing or multimedia messaging on handsets.

Mostly the automobile manufacturers would build a system of computers, storage devices and communication devices, where they could rather well predict what would be the typical traffic which a typical intelligent car would be placing on to the network. This traffic would balance out over the thousands of cars that a manufacturer would sell in any given year. That total traffic would then be offered to the 3G service providers on a wholesale basis for the full fleet of cars, and on an annual contract basis. The individual driver would not negotiate with the network providers, but rather the communication package would be part of a 'Golden Care' or some premium service package offered by BMW or Mercedes Benz or Audi, Jaguar, Volvo, Toyota, Cadillac, etc.

Hint to application developers: follow what is happening in the car industry. The automobile industry has huge budgets and faces strong competition, manufacturers will be major innovators in the telematics area, and other application developers can learn a lot from them.

As the major manufacturers would be negotiating for a large volume of traffic for a large amount of cars, they would have considerable negotiating leverage. And as the data traffic generated by the cars themselves would be very low, the total traffic and the total revenues from it would also be relatively slight for the overall big picture of 3G revenues and traffic. Of course the phone calls placed from cars, and the possibility to surf the Internet while someone is driving the car, will provide voice and data access, but these will be similar to those discussed elsewhere in this book.

The automobile service garages industry and the car customising industry are a big part of most economies and will look to develop services around these innovations. Most of these will be building 3G solutions into existing cars for current owners who want the utility but don't want to buy a new car just because of this one benefit. The companies well poised to capitalise on these are those specialist garages who now install car stereo systems, hands-free mobile phone set-ups, and car alarm and disabling systems. They will be needing support and advice from equipment manufacturers and from 3G network operators. Probably in most countries there will emerge one or two preferred 3G network operators which will use these independent garages as a delivery channel to access the independent car owner.

10.2 Remote views

Another major area of automation and machine communication is the ability to access fixed location video cameras and view their transmitted images on the 3G phone. It will also be possible to have the mobile handset control some of these cameras. Many services can be built around the ability to see in a remote way.

10.2.1 Remote eyes

The Show Me type services will work well when there is another person with a 3G phone who can point the camera on the phone towards whatever needs to be seen, for example, items to select at a store. Similar uses will become commonplace with remote control cameras, some of which will have universal access, others will be private 'remote eyes'. A typical science fiction example is the remote control camera inside the refrigerator. The 'Fridge Cam' will allow you to call your fridge and use your 3G phone to see if there is milk or eggs, etc., and decide what you need to buy.

The fridge cam is likely to be a rare use of 3G technology, but more common will be access to home cameras for checking in with the children for example that they are doing their homework. Typically the service would be set up so that the home owner has a private access number to the home cameras (with considerable security so that burglars cannot tap into this line). A similar set-up could be where the home camera calls you, to show you if something is happening, for example if the cat is having a fit and starts to tear at the curtains. There can be triggers of the level of uncommon activity which could trigger an automatic call. You would get a call from your home, showing you what is going on.

Soon these types of devices will be automatic enough that you can just stop by at the electronics store to buy a cheap remote camera. It can be plugged into the fixed network, such as the phone line much like adding an answering machine, or the cable TV connection, like adding a cable TV 'box' – or they can be configured to use the 3G network, or other wireless technologies. Then there will be a simple connecting protocol so that your private 'remote cam' will call your 3G phone(s) and create the necessary security link so that you can access the camera. In a solution like this, the automation will need to be created by the camera and its software, but for the 3G network operator, the video image will be only a data stream no different from sending a file transfer when you connect your computer to the network or send an e-mail with an attachment.

10.2.2 Show the city

Numerous live cameras can be set up around the town to give live video feeds of traffic congestion points and parking lots, etc. Such cameras could also be used to promote lively nightclubs and pubs to show what kind of crowds are at the nightspots, etc. Or in an opposite marketing method, places with traditionally long lines and waiting periods and big crowds, such as amusement parks, could show via cameras when they have little crowds to entice more visitors.

With these types of cameras there would be fixed Internet-enabled access to these public cameras, and there would be a connecting charge most probably billed by the number of seconds you actually send the video feed from the remote camera to your 3G phone. When there would be mobile video access, such as temporary video feeds from special events, these would of course cost more than the stationary cameras. But mobile remote access could be sponsored by advertising, for example, the views to lively nightclubs could be sponsored by a brand of beer, or the traffic congestion cameras could be promoted by a chain of gasoline service stations.

As the screen size of the terminal increases other options have to be found for the keypad. All phones will not be based on PDA-pen input methods and voice recognition although improving is not yet suitable for the mass market. One option to get around this problem is demonstrated with a variation on terminal design this time from Hiphop.

10.2.3 Military use of views

There have been many isolated cases of modern warfare using cellular phones to transmit a given communication, as well as paramilitary action such as to organise riots, etc. A whole separate consideration is the military aspect of 3G cellular technology. Compared with military specification robustness and reliability, as well as the ability to withstand weather and punishment, individual components of the traditional 3G phone and network are definitely so fragile and vulnerable as to be almost useless to construct meaningful military control *during wartime*.

But during peacetime, the 3G network, terminals and applications provide a vast array of communication and surveillance technology which the military can use for training and for coordination and logistics uses. Again, a whole military manual could be written of what all could be done to improve peacetime military communications, but suffice it to say at this point that apart from voice and click-to-talk, the 3G network could bring in the benefits of instant messaging, chat, show me and live video, and IP-based intelligent devices which could communicate automatically with each other. For example simple cheap remote surveillance camera-phones could be used to monitor the progress of a battle exercise. Other remote devices could be used to allow remote detonation of dummy charges, etc. Of course any battlefield robots and RPVs (Remotely Piloted Vehicles) could be guided in peacetime with 3G network-based technologies, probably to great savings over the much more secure military spec technologies needed during real war situations. When combined with possible synergies for example with the police forces, solutions can be built at very reasonable costs.

10.3 Remote control

If an automated device can have the eyes of a video camera, and remote access to control that camera such as its zoom or panning mechanisms, then it is almost no step beyond that to include remote control of more than just the camera. The 3G devices will evolve soon to include a multitude of specialist devices which include some 3G terminal ability, and special application software to allow remote control of the whole device. Often this could include the incorporation of the video camera so that the remotely controlling person can 'see' what needs to be done in remote control.

10.3.1 Home control

To see something remotely is only one small part of the 3G opportunity. As 3G allows for IP traffic, it is quite feasible to construct remote access services, where one can remotely control electronics and robotics. Already many simple solutions exist by which one can remotely adjust the temperature of one's home, so that it heats up a little before you get home – important in cold climates – or the air conditioning cools the house just before you get home – important in hot climates. Also controlling lighting, alarm systems, garage doors, even setting up the sauna to heat up, can be controlled with current technology and a mobile phone.

10.3.2 Home alarm systems

Another variation of the remote control theme is the home alarm – or office alarm/industrial alarm – for that matter. Whether it is a fire alarm or burglar alarm or water damage, the ability to send the notice immediately to the mobile phone of the person or company owning the property, is potentially a great value to the owner of the property. This is one extreme example of where the cost of delivering the information – the alarm notification can be sent in a package of a few bytes – and the value of the information being received 'your home fire alarm has just been set off' is usually dramatic. The natural bundle to this service is remote control cameras, so that you can also take a look and see if it was only the neighbour's dog which set off the alarm.

10.3.3 Household gadget control

The expansion of such opportunities is only limited by imagination. Of course things such as remote setting of the VCR will be easy to do via the 3G mobile phone, or the turning off of the family TVs when the young children are supposed to go to bed, equally so.

The money in remote control would be more around the technology integrator which creates the remote control solution. Home security companies, such as the ones providing home alarms and guarding services, are naturals to be strong early on.

10.3.4 Industrial metering

Another obvious early use is to use electronic sensors and 3G networks to have industrial meters reporting volumes of usage, from electricity, gas and water to the home, to matters such as gas pipe pressures on gas pipelines, and water surface heights at controlled rivers and reservoirs, etc. etc. etc. In many cases the service could be installed also on a fixed line, but a dedicated cellular data connection and terminal device can be more flexibly adopted and connected in a universal way, than pulling telephone cables to various metering locations. When the device does not need to have telephony ability, nor the ability to dial more than one number, then the 3G remote metering device can be built very simply and inexpensively.

> **Hint to 3G network operator: if government is a big customer, get into metering.** If your target customer includes the government agencies and related organisations, then understand that the utilities sector will be going heavily into remote metering and control. Get competent in it.

For the cellular operator, the individual metering devices are excellent customers. They do not put a lot of traffic on to the network, so it is almost impossible for the machine-to-machine traffic to overload the network. The traffic is typically a short duration and to dedicated designations, so also predicting the traffic is easy. The service is almost never 'real-time' time sensitive – for example the electric meter needs to be read once a month, and it does not matter if the metering result is transmitted on the exact minute it was read, or 25 min later, or even during the following night. So machine-to-machine traffic is typically not going to present network dimensioning challenges.

10.3.5 Inspections

A common need in business and government is to inspect almost anything. Often the inspection work is routine, but the exceptions to the inspection need the seniority and know-how of the inspector. 3G remote viewing, remote control, and robotics allow the creation of various intelligent devices to do part of the inspecting, so that knowledgeable experts can be used only when needed. In random cases of 'what is going on' it will also be possible to send a junior employee with a 3G phone, and if the manager needs to 'see' what is happening, the junior employee can show it with the camera on the 3G phone.

In most inspection solutions the integration of the specialised instruments and equipment with the 3G network will need IT system integration ability. 3G network operators typically do not possess that know-how, at least not in their mobile network business. So this is an area where cooperation with the IT sector is likely to be needed. The possibilities that the technological innovation will make many currently employed inspectors and line managers redundant, brings about the need for organisational and human relations consulting, which in turn would introduce management consulting needs into building a working solution. Any 3G service would need to be designed to accommodate the contribution and value of these players. If the proposed service involves inspections of objects which move a lot – such as cars on a train – or inspectors who move a lot, then the mobile network is probably the only real platform for which a total solution can be built. This ensures that a solution cannot be built without the 3G network operator's involvement.

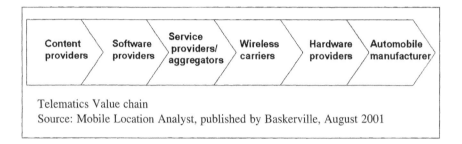

Telematics Value chain
Source: Mobile Location Analyst, published by Baskerville, August 2001

10.4 Law enforcement

Police and law enforcement are deploying ever more technology to catch people who break the law. One area is drivers who break speed limits. Currently police have radar guns, unmarked cars with calibrated speed-ometers and cameras, helicopters, and automated sensors. Soon mobile Inter-net applications will bring new utility to this branch of government.

10.4.1 Speed trap

The automated speed trap will have several possible ways it can evolve. One is the simple remote mobile camera connected to a radar gun. A remote location automatic radar gun could measure the speeds of all incoming auto-mobiles, and a camera could then log the picture of the car, driver and store the record of the speed measured by the radar gun.

Another development is the always-on connectedness of 3G. Soon the car will be continuously connected to the 3G network. It will become very easy to deliver alarms, messages, warnings – and notices of speeding tickets – to drivers as they drive. Very rudimentary applications of this type of direct messages to those who break the law have been used for example in the Netherlands where police send messages to stolen mobile phones requesting them to be returned to the police.

The speed trap would have very different traffic patterns from a human. The speed camera would be uploading constantly images of speeding cars, never needing to take a break to eat or sleep, and in that way would probably load the network more than the most avid web surfer could hope to match. Yet from the police force point of view, the small costs of the picture transmissions would probably be significantly less than the costs of manned speed traps rushing after the speeding cars.

10.4.2 Talk to criminal

Another area of law enforcement would be the ability to communicate with criminals as they are in the action. With the proliferation of inexpensive 3G cameras and other monitoring devices, soon there will be the chances to record the activities of criminals from a multitude of cameras while they are starting to commit the crime. The police would be summoned and at some point, when the police were close enough to be sure of capturing the criminal, the police could start to talk to the criminal. For example this could diminish the amount of damage being done to a home or car. As the remote cameras would feed 3G images to a network server, there would be no benefit in trying to destroy video surveillance tapes or recorders. It would not put an end to crime, but in many cases talking to the criminal would help reduce the total damage done by the criminal.

10.5 Feedback systems

An area which is likely to grow fast is that of feedback systems. As worldwide competition becomes ever more intense, the ability to react to customer opinions will become equally ever more critical. The electronic feedback ability of the Internet has given many players there a remarkably strong tool to harness customer opinion and develop the overall customer appeal of their service. Now it is of course quite normal to have instant e-mail reply ability at most websites. But just like with SMS (Short Message Service) text messaging being far superior to e-mail in the *speed* of delivering messages, so too will 3G response systems be able to bring the Internet speed up to yet another faster level.

10.5.1 Feedback-based control

At such speeds feedback solutions cannot always wait for managerial review. There will be many places where the feedback can be automated. For example the feedback on the opinion of the temperature at a conference centre. There is no need for staff to notice that several conference centre visitors are complaining about the heat before they turn up the air conditioning, there could be an automatic address, and if anybody had a complaint about the heat, they could send the complaint directly to that control. The system could be automated so that it would balance the needs of received replies and average the requirement, even reply automatically to the persons complaining. So for the first person, the system could say that it thanks them for the input and sees what it can do. Over the next few minutes as a few other people 'complain' the system could initiate a temperature change, and let the complainers know. If then, before the change takes effect, there was another person complaining again, the system could say that the matter has already been addressed and that the temperature change is already taking place.

10.5.2 Customer surveys

Automatic feedback systems could be used for most kinds of situations from returning your rental car to opinions on how the teacher taught the lecture today. The system could easily provide filtered feedback to the persons involved so that exceptional comments are excluded, such as using profanity, etc.

The money in feedback systems lies partially with the system integrator, but to a large degree with the 3G network operator, as it can make much use of its databases and provide for example customer profile data without revealing individual information. This could be for example the male/female mix of replies, the age distribution, the top five countries where the main visitors were from, etc. If the customer survey asks for this type of information, it is likely that the respondent will not want to complete it, or give incomplete and faulty information. The 3G network operator can easily give summary data without revealing any customer sensitive information. But the 3G network operator would not give any information for free.

10.6 Other machine services

It would be impossible to even outline all possible uses of automation in information flow and control, and their benefits. Just like the computer and

PC have become commonplace everywhere from the dentist's office to the hotel reception, so too will 3G automated communicating and computing devices. It is important to remember that these often are not traditional phones, but can be very customised devices which have the ability to communicate via the 3G network. The whole solution can easily include many components which connect in fixed or other mobile ways. For example a security solution could be fixed in the building, but the security guard could inspect the status via his 3G terminal. The UMTS Forum in its Report 16 discussed the machine population showing how the machine population is expected to grow over the next years.

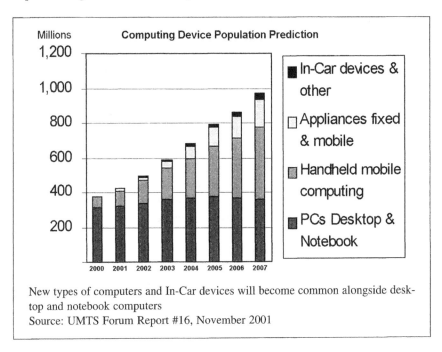

New types of computers and In-Car devices will become common alongside desktop and notebook computers
Source: UMTS Forum Report #16, November 2001

10.6.1 Traffic monitor

Various cities are experimenting with ways to automate the collection of real-time traffic data. In most Western cities the local news will give traffic congestion updates on the radio, and these are even collected by flying traffic monitoring helicopters. A much cheaper and more accurate real-time system can be built around remote monitoring devices feeding information as needed to a centralised traffic status system. Messages from that system could then be sent as requested or subscribed, to motorists, the police department, taxi services, etc. Automobile navigation systems and in the future car auto-pilot 'chauffeurs' would be natural clients for this type of information. The machine-to-machine communications would vary depending on automobile traffic and weather, etc. conditions.

10.6.2 Automated hotel receptionist

One of the near future visions that has been suggested by many, is the integration of the frequent traveller information in hotels and airlines, together with the automated booking systems, and the credit card systems, and the hotel receptionist system. This system, combined with sensors could result in a traveller-friendly hotel entry which could go like this. You have selected the hotel on your 3G phone some time ago. The system knows your preferences of non-smoking room and high floor with view. As you have been driving your rental car near to the hotel, the network noticed that you were near the hotel and alerted the hotel. As you step out of the rental car the hotel system recognises you, welcomes you, and lets you know which room has been assigned to you and on what floor.

The hotel system knows already your preferred payment method, and you are already aware of the way to use the mobile phone as the room key. As you approach your room door, the Bluetooth-enabled door lock identifies your phone and opens the door. Of course the whole procedure tracks what payments you may have incurred, such as the parking charge, etc., and collects these on to your hotel bill. A system like this would be built over time, and would require that most frequent travellers to that hotel have 3G and Bluetooth-enabled phones, but that vision is not far in the future, and as this type of automation greatly saves in the costs of running the hotel – the personnel costs involved in handling the check-in and payments – the hotel industry has great incentive to develop their systems in this direction.

10.6.3 Automated airline check-in

A similar application is the automated airline check-in. There is little reason for a frequent traveller to actually meet a check-in officer of an airline at the check-in counter. Much of the preferences are stored in the system, and the traveller would be much more rapidly served via self-service. The first such systems are coming on line during 2002.

10.6.4 Software upgrades

A few words have to be said about the inherent ability of the 3G network to deliver software. The 3G network, the terminals and mobile phones, and the applications used to run 3G services, are all naturally able to be upgraded via the network. The limits to this tend to be with the manufacturers who may want to limit or control the extent. The various upgrades can be offered, sold, delivered and paid for via the 3G network. The various upgrades to the games, browsers, applications, operating systems, on down the line to the software in the network, and on to the billing systems, etc., should all be done via the 3G network. Such upgrades embody the very essence of the 5 M's, with upgrades recognising the Movement, the Moment, the Me, the Money and delivered of course through Machines.

The decisions on which parts of the end-to-end solution are best to upgrade, and what means are best to do so will differ by service, application and device, as well as how the 3G network operator wants to develop its enabling systems. But these upgrades will naturally happen through the network. The ability of the 3G network operator to provide security to the transfer of the software will provide considerable utility to the software developer. Software developers are continuously concerned about intellectual property rights and any infractions against them through software piracy and hacking. While of course not a perfectly secure delivery mechanism, the 3G network is designed with numerous features which provide strong security to the communication.

10.7 Real services today around the Machines attribute

Again, the ability of machines to communicate with us and with each other is nothing dramatically new, simple solutions have been appearing around the world. Here are a few of the more familiar ones.

10.7.1 Driving aid systems in cars

The intelligent car concept with maps, spoken guidance, location information, entertainment, etc., is already in full deployment in many countries. Some of the pioneers have been the Detroit car manufacturers, and some of the positioning data, etc. is fed by the GPS system. Early user feedback is very positive as well. Most major car manufacturers have announced ambitious schedules to deploy navigation, security, safety, service and entertainment systems.

10.7.2 Mobile car security

In the US several car security related solutions already exist, ranging from the wireless operation of the car door locks, to tracking and locating a stolen car, to remotely accessing its systems.

10.7.3 Intelligent tyres

The evolution of intelligence in and around the automobile has reached what used to be a rubber item – the car tyre. A tyre manufacturer in Finland has announced a Bluetooth-equipped tyre which will monitor itself, and let the car (driver) know if there is a problem with the tyre.

10.7.4 Gas, water, etc. metering

Some of the early remote metering solutions have been introduced by various government utilities in Britain, Sweden and Australia to remotely read water, gas, electricity and other such usage.

10.7.5 Home remote monitoring and control

The intelligent home is being developed in many countries. Some of the early remote access and control solutions have been introduced in Germany and Finland, where you can set home lights, temperatures, even turn on the sauna remotely.

10.7.6 Software upgrades to mobile phones

Some mobile phone manufacturers have already introduced ways to get software upgrades to their systems through the mobile network. One of the pioneers in this was Nokia with its Club Nokia concept.

10.7.7 Appliance control

Some appliances and devices are already in existence which can be controlled via the fixed Internet. For example VCRs, washing machines, refrigerators exist which allow control via the Internet.

10.7.8 Police messages to criminals

Some of the first trials to have police communicate directly with criminals have been of the exceptional nature and reported in the press. Perhaps the best application is in the Netherlands where if someone has a stolen mobile phone, the police start to send messages to that phone at short intervals saying that the phone is stolen, and it should be returned to the police.

10.7.9 Automated voice systems

In the US several voice message delivery and voice broadcasting systems have been developed and politicians have found that short political messages delivered via voice broadcast to telephones is an effective way to enhance a campaign. The voice rich services tend to be focused on fixed phones but are likely to soon arrive also on to mobile phones.

10.7.10 Customer surveys by mobile

In Sweden the market research industry and mobile telecommunications have introduced the use of SMS and WAP (Wireless Application Protocol) into marketing research such as opinion polls and surveys, etc. The industry has found it a great saving in completing the research and accuracy in the digital data produced.

10.8 Blockbuster

This chapter has examined the revenues and profits of Machine services. In-car service is one of the fastest growing areas for car manufacturers as well as wireless and non-wireless service providers. There are a great deal of expectations and much hype for next generation solutions and only time will tell whether these expectations will be met. The 3G era is about services and the car is a very good example of the new possibilities in terms of environment, technologies and applications that service providers are facing. New entrants into the mobile service opportunity, in this case car manufacturers, are planning to challenge the existing service providers in different vertical segments. Whether different vertical markets are served by one company with a broad portfolio or number of companies with more focused product portfolios – or by a combination of these two – is also a very interesting question.

Less glamorous Machines services will also appear from monitoring to controlling various devices and gadgets. The Machine population is likely to exceed the human population in 3G subscriptions before the decade is done and the revenues and profits of these services will play a key role in determining the business viability of the whole industry. Regardless, with Machines services just like all new mobile services, it is more important to be able to quickly adapt to changing circumstances than it is to be perfectly right to begin with. Perhaps Steven Wright can remind us of how different the opportunities can be for the players with his statement, ''The early bird gets the worm, but the second mouse gets the cheese.''

11

'A page of history is worth a volume of logic.'
Oliver Wendel Holmes

Money Patterns in Cellular Networks:

The 'hockey stick' curves

Traffic patterns in cellular networks take on a particular form. Network costs in cellular networks also take on a particular form. These patterns differ from other more established patterns in telecoms and data networks. This chapter examines briefly why traffic in cellular networks distorts Metcalfe's Law and produces a more pronounced inflection point – resulting in a pattern identified as the hockey stick curve. This chapter also looks at how cellular network costs occur over time, which results in an inverted hockey stick pattern. Finally this chapter combines these two phenomena on to one time scale into a theory called the 'Hockey Sticks' theory.

11.1 Cellular is a distorted case of Metcalfe's Law

The growth patterns in traditional voice services on fixed telephone networks are at or near saturation – in some countries already diminishing having peaked. Fixed telephone network traffic patterns are very predictable and tend to follow near-linear patterns. That is because the total number of connected voice callers has plateaued. With fixed telephone networks the utility and traffic patterns follow Metcalfe's Law – the utility of the network increases with the square of the number of connected parties. (Robert Metcalfe founded the computer networking company 3Com.) Usually Metcalfe's Law is expressed as an exponential curve as is shown in the following figure.

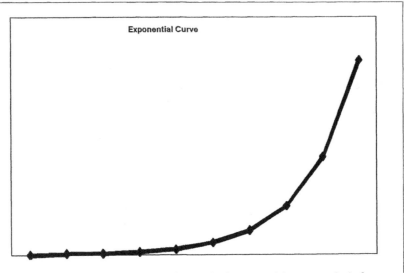

Exponential Curve

The standard Metcalfe's law produced a standard exponential curve, typical of most telecoms services

Metcalfe's Law is based on the assumption that the value of the service is uniform to connected parties and Metcalfe's Law has been well documented and accepted as explaining network traffic in fixed telecoms and fixed datacoms networks.

11.1.1 Hockey stick is not Metcalfe's Law

The traffic in cellular networks initially seems to follow Metcalfe's Law until a certain 'critical mass' has been achieved, and then the cellular network traffic pattern breaks that which is predicted by Metcalfe's Law. Cellular networks have a much more pronounced change in the traffic, producing a clear inflection point. The inflection point is dependent on subscriber numbers. The phenomenon was first noticed on GSM (Global System for Mobile Communications) networks when the first countries started to experience high 20% and low 30% penetration rates. When this happened, those isolated incidents were dismissed as anomalies in the various early adopter countries of GSM. But when the pattern repeated itself across all cellular networks for voice, and started to repeat itself in the same markets for SMS (Short Message Service) text messaging, it was obvious that Metcalfe's Law was not directly applicable to cellular networks.

Individual operators could not verify that this phenomenon occurred in other networks, so the first to notice the pattern was the network planning engineers at the equipment manufacturers. Cellular network engineers with their network equipment manufacturers mostly situated in countries like Sweden, Finland, Canada, Germany, etc., where ice hockey is also quite popular, very quickly named this phenomenon the hockey stick curve. Some people from other networking technologies have then proceeded to call all fast growing usage curves 'hockey stick curves'. In most fixed network cases the incidence is actually normal Metcalfe's Law and to call such patterns hockey sticks is not accurate. Most fixed networks conform to Metcalfe's Law at its basic form, in an exponential benefit curve. Cellular networks do not. They may be a special case of Metcalfe's Law but the shape is distinctly not exponential, it is more pronounced. Cellular networks are somehow a special case and further understanding was needed.

It is now clear that cellular networks introduced a new value element to the cellular network, which does not exist on fixed telecoms and fixed datacoms networks. That new benefit is called 'Reachability'. This produces a distorted Metcalfe's effect, more accurately portrayed by the form of a hockey stick curve. Thus network utilisation in cellular networks does not fit the traditional exponential curve as predicted by Metcalfe's Law, but rather cellular network traffic and revenue have a more pronounced inflection point and the hockey stick effect is more dramatic.

11.2 Phenomenon of Reachability

Reachability is a new phenomenon which is still only understood in its earliest levels. Reachability has already had a profound and unanticipated positive effect on cellular networks, and is starting to cause change in behaviour overall in society. Reachability means the ability to be contacted. Reachability may seem like an obvious generality, which it is not. It is a fundamental new effect and it will modify human behaviour significantly over the next decade or so. Its effect on mobile telecommunications is best explained by a simple example.

11.2.1 Honey I'm a little bit late – spontaneous contact

Reachability is best explained by the example of how being a little bit delayed has changed over the past 10 years. Ten years ago almost nobody had mobile phones. So if the husband was running 10 min late in picking up the wife, he would not stop the car and look for a telephone booth to place the call. It was not worth 10 min, and the delay in parking the car and making the call would double the delay. So there was no call, he would simply arrive late. So even though there was theoretically a need to call the wife, the husband would not use the fixed network to act upon that need. So this pent-up demand is not met by the fixed network telecoms traffic. The need is subconscious in that most people would not even think of wanting a mobile phone to meet that need.

During the past decade the husband more likely was the first to get the family's first mobile phone – very possibly as a phone from his employer. One of the new effects was, that if he happened to be late to pick up the wife from home, he could now call from the car. This is the direct effect of tele-communications fitting the normal Metcalfe's Law and illustrated the husband's behaviour of placing the call. There is no reachability yet as the wife is home where there already was a fixed phone. For the husband this – calling from the car when a little bit delayed – is not an immediate change in behaviour. Some people notice it almost immediately once they start to use a mobile phone, others take much longer to even 'remember' that they have their phone and they could actually call to let the waiting party know. Regardless, the spontaneous need to make contact is real, it is enabled by mobile phones; it was not recognised in early predictions about cellular telephony. It was unknown at the time, an unmet, pent-up, spontaneous need.

11.3 Reachability

The second early inhibitor was even more difficult to predict, and only became easy to understand after people had started to use their mobile phones and keep them on. It is the unmet pent-up need to be able to be contacted, to remain connected, of being 'Reachable'. Reachability arises only when the receiving party has a mobile phone (and keeps it turned on). Let's start by setting up a problem in the previous example. The husband is again late, but this time he is supposed to pick up the wife from the shopping centre. While the husband now has the mobile phone in his car and could place the call, but with the wife not at a fixed phone, he cannot reach her. While the husband could make the call, because she does not have one, he cannot complete the call. So he decides to buy her a mobile phone for Christmas. Next time the husband is late, she has a mobile phone in her purse, and now the husband can tell the wife that he is delayed. What Reachability allowed, is the short update calls to let the waiting party know that there has been a change, or that one is coming, etc. Reachability is a radical change in mobile telecoms from older telecoms communications. Only with Reachability do we get the short informing calls of updates of where we are.

> Hint to network operators. Understand Reachability. It is a new phenomenon unique to cellular networks and started to be discussed only at the end of the 1990s. There was no literature on it when your staff went to university. But Reachability is a competitive advantage to you. Every R&D, Product Management, Marketing and Sales person should fully understand Reachability.

11.3.1 Reachability and the call-receiving person

Reachability changes behaviour in the call-receiving person. Early on many complain about the mobile phone, say they will not bother to carry it with them, and see no need to keep it handy at all times. It is only through discovering Reachability that the mobile phone becomes indispensable. As the person notices that the mobile phone delivers critical updates of what is going on, the person starts to value keeping it always handy, and always on. It takes numerous individual instances of reinforced behaviour to produce change in behaviour, so the events that involve benefits from Reachability will take time to cause societal change of behaviour. The changes are very clear in countries in Scandinavia for example, and many Western European and Asian countries are following suit.

11.3.2 Reachability and the calling person

Reachability results in change also in the calling person. With a fixed tele-phone number, you call a *place*; with a mobile phone number – you call a *person*. With fixed calls you call an office or home, and hope that the party you want to reach is there. When calling a mobile phone you assume that the only person who could answer that phone is your intended party.

11.3.3 Reachabilty and messaging

The same phenomenon of Reachability can be observed in mobile messaging. With e-mail on the fixed Internet, you send the e-mail and you assume that the person will read it when the receiving person next is at his or her e-mail; i.e. if sent to a work e-mail address, you assume it will be read during office hours

11 Vignettes from a 3G Future: m-Forms at Work

I hate the paperwork that seems to be needed for anything at work. But at least the forms have become more intelligent, and now we can fill them out by 3G phone. The travel expense form pops up automatically whenever I return from a business trip back to the home airport, and it has saved me many times. Now that the accounting department at the office has finally accepted the still images taken with the 3G camera of the receipts, the whole expense form process can be handled during the time the airline serves coffee.

B2E (Business to Employee) systems benefit greatly from e-business electronic form filling, from savings in paper printing, to mistakes in forms, to processing of forms to the electronic trails of who filled what and when, to management reports from the data. The m-business application of m-forms will be a natural progression of this, bringing the form filling even closer to the actual occasion or need which results in the form. The tight integration with the company's B2E systems is essential, but the savings in form processing alone can very easily justify the migration of employee populations to new mobile phones.

and if sent to a private e-mail, probably read during the evening or weekend, etc. But with SMS text messaging the message is assumed to be with your intended party within seconds, and read within minutes of being sent. Else you assume that the person is unable to view the message such as being in a meeting or in an aeroplane, etc. Again you reach the person much more consistently and considerably faster using mobile datacoms than fixed datacoms. Same factor. Reachability. It too has brought about radical change in behaviour, such as making last minute plans and changing them on the go. And of course, SMS text messaging follows the hockey stick curve, specifically because of Reachability.

11.3.4 Reachability and America

This is also a critical lesson to American readers, who tend to have 'receiving party pays' type of cellular phone billing. Receiving Party Pays is very counterproductive to cellular phone traffic, as people tend to keep their cell phones turned off. For non-American readers it may be of interest, that American cellular phone users often keep their phone turned off, and carry a pager ('beeper'). If you want a friend to accept your call, you first send a message to the beeper, and then call the person. This Receiving Party Pays system strongly diminishes Reachability, possibly even eliminating the chance for Reachability to be discovered by the cellular phone user. It is probably the biggest single reason why American cellular carriers are not seeing the same explosive rate of growth in users, traffic, revenues and profits, as their European and Asian counterparts. American carriers should very aggressively pursue 'calling party pays' type of cellular billing, the sooner the better for the whole industry. If a phone user knows that he never has to pay for someone calling his phone, the user will keep the phone turned on. This allows him to discover Reachability, and become much more addicted to keeping his phone on, and then using his phone at all times in all places.

The nature of mobile phones is that they can allow spontaneous attempts to make contact, and also enable the phenomenon of reachability. Before the introduction of mobile phone services, both of these factors were unknown, pent-up, subconscious needs.

The phenomenon of the hockey stick curve is still relatively young – the first GSM networks only appeared 10 years ago, which were the first cellular networks with two significantly different services – voice and SMS text messaging. As the concept of the hockey stick curve is still so young, its reasons are not fully explained, but some assumptions have been made about what causes the abrupt inflection in the service usage.

This reachability, combined with the spontaneous ability to make contact, is what makes the mobile phone so addictive as a personal communication device. It also means that once people have the phone with them at most times – and importantly when considering the American marketplace – as long as they keep the phone on at all times then the people will be receiving more calls. The reachability aspect of mobile telephony was totally unknown to the industry, and for individual users it was a subconscious pent-up demand. Reachability became known only after people started to use mobile phones.

Both phenomena of spontaneous ability to make contact and Reachability, mean that there is the very real, but early on unknown, need waiting to be met by cellular network services. As the users themselves do not know these needs, and are exposed to them gradually and accidentally, the traffic does not illustrate its true nature early on. Not until a critical mass of users have learned the extent of needs they can meet with the new services, the traffic pattern lags behind the true levels resulting from the population and its needs.

11.3.5 Reachability as competitive advantage

Reachability is unique to cellular networks. It applies to all human communications using mobile phones and other mobile devices. Reachability is a totally new phenomenon, mostly only discovered even conceptually in most markets at the end of the 1990s. Reachability is already causing dramatic change in human behaviour, and this without any active and purposeful intent by the mobile operators and service creation experts. The first lesson to any reader of this book is to be sure to understand what is Reachability and how it impacts the user, through creating addiction, binding the user to the mobile phone, and through valuing those services which deliver or enable Reachability. No technical competitor which cannot deliver Reachability will be able to supplant a mobile phone once its user has discovered Reachability.

The mobile operator needs to ensure that all users discover Reachability and learn to use it. Various services, such as voice mail and SMS text messaging can be used to train users. Also campaigns, with themes like 'Are you running late? Send an SMS', etc. will not only increase traffic, but teach the users to value Reachability. Some people learn fast, for others it may take years until they are hooked.

More importantly, the mobile operator needs to deploy solutions and services which capitalise on Reachability. With this phenomenon, the mobile operator is uniquely able to monetise it and create value and profits out of

Reachability. Any service which builds upon Reachability will have a natural competitive advantage against all copy-cat services that rely on fixed networks. Of course Reachability can be used from fixed networks as well, to reach someone on a mobile phone. But at least initially, this is a natural strength of mobile operators and thus early service ideas using Reachability should be built by them.

11.4 Service usage pattern: hockey stick curve

As we saw, the usage and traffic patterns in cellular networks have a more pronounced inflection point than the gradual exponential growth as predicted by Metcalfe's Law. The hockey stick curve has two distinct parts. The first part is a gradual growth path when the mobile service is used to substitute for known fixed network services to meet familiar needs. The second part happens at an inflection point in time, when a critical mass has learned to use the services also in new ways. At that point the angle of the usage curve takes a decidedly more steep angle. Usage starts to increase dramatically more than before.

This results in a particularly recognisable shape of what is often called the 'hockey stick' curve. This pattern has emerged in most cases of cellular voice service use, and practically all SMS service use.

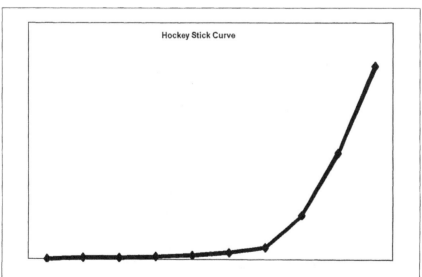

Mobile services adoption takes the form of a more pronounced inflection point in what is called the 'hockey stick' curve

The 'hockey stick curve' is a relatively broadly accepted pattern of how service usage – and traffic patterns and thus also usually revenue patterns – develop in mobile services on cellular networks. With cellular voice the inflection point has been found at about 17% penetration. With SMS text messaging, when networks allow SMS roaming across networks in the country, the inflection point arrives at about 28% penetration. Other services are too young to have enough data to isolate the inflection point.

While Metcalfe's Law and the Hockey Stick Curve predict traffic patterns, one can draw revenue assumptions from it. If we assume that the cost per minute (or message or megabyte) remain near-constant over time, then the Hockey Stick Curve will also relate (near) directly with revenues of the same pattern.

11.5 Network build-up pattern

Cellular networks are built in two phases. For traffic to be possible at all, initially radio coverage must be achieved over a given geographical area. Once coverage has been achieved, then most likely there will be pockets of heavier traffic – such as cities. The nature of cellular radio transmission is that there is a theoretical maximum of simultaneous connections that can be managed within any given 'cell' of radio transmission equipment. A cell holds typically one base station and one antenna. This is not a book on how to build and dimension a cellular network, and for more information on how a UMTS (Universal Mobile Telecommunications Services) network is built, please see Kaaranen *et al.*'s book *UMTS Networks*. Kaaranen *et al.* offer a simplified conceptual illustration of how cellular networks initially are built.

Initially cellular networks are built to cover a given geographic area. It is the nature of cellular networks that cells can be made smaller, in other words a cell can be split into smaller cells. This increases capacity but adds cells and thus cost. Those cells can be divided into smaller and smaller cells, which then again can handle approximately the same amount of simultaneous calls each, which the single cell was able to do earlier. In this way there is a lot of theoretical growth possibility in the capacity of a cellular network, simply by increasing the number of cells – and thus increasing the density of the antennae and base stations, etc. Of course there are technical limitations to the absolute density, and other factors also matter in that the cell sizes cannot be forever 'split ever smaller'.

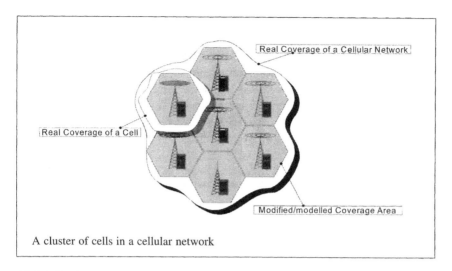

A cluster of cells in a cellular network

11.5.1 Upside down hockey curve

The build-up of cellular networks has thus two distinct phases – the first phase is totally predictable and relatively uniform in density: it is the build-up of coverage. The operator/carrier will have a certain demographic or geographic coverage target, and it will build enough cells to its network to achieve that target. There are a somewhat standard size of cells for rural areas and urban areas, and depending on national legislation on antenna heights, radiation levels, etc., initial cell sizes can be defined and a radio coverage initially planned. When there are only a few users early in the network, there are no areas with capacity problems, and the 'thin' geographic coverage is adequate to support all callers. The initial build-up is predictable in size and tends to be rapidly built.

The second phase is the additional build-up due to capacity needs. This will happen in concentrated areas of traffic, typically where people work and live in close proximity, urban areas of cities, etc., as well as places where large numbers of people collect together, such as sports stadiums, etc. Throughout the network there will emerge areas where any given cell cannot support the number of callers that typically place or receive calls within that area. That cell is then divided. This type of growth is more gradual than the initial build-up phase, and will be reactive to recorded network congestion. When plotted over time, the initial heavy build-up for coverage, and then the more gradual continued build-up for capacity, produces a shape which can be called an 'inverse hockey stick' as can be seen in the following figure.

This pattern of cellular network build-up rates and their related costs is the same regardless of technology. Thus the same inverse hockey stick would apply in any case of cellular networks, be it CDMA (Code Division Multiple Access), TDMA (Time Division Multiple Access), GSM, WCDMA (Wideband CDMA), CDMA2000, etc. It is also something which is not well understood among telecoms industries from the fixed side, where the build-up costs are more accurately related to capacity needs. In cellular networks capacity is eventually a driving factor, but coverage is the initial factor.

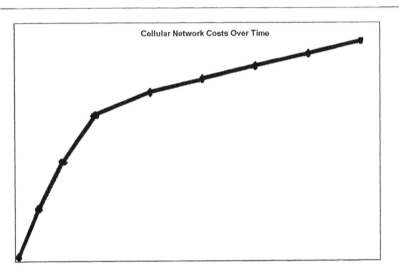

Cellular Network Costs Over Time

The costs of building up a cellular network have an early high cost part and a later gradual cost part, resulting in an 'inverse hockey stick'

11.6 Double Hockey Sticks

There is a relationship between how networks are built, and how services are used – or from a business case point of view, a parallel relationship on how the network build-up costs relate to the service revenues. This relationship was first explored by Matt Wisk of Nokia when he was developing the theory he calls the 'Double Hockey Sticks' with its 'Window of Worry'. I have worked with Matt Wisk to refine that theory to incorporate details in the timings, etc., but as the father of this theory is Matt Wisk, I like to consider this the Wisk–Ahonen theory of Double Hockey Sticks.

The Double Hockey Sticks theory shows the relationship over time of network build-up costs, and of that network's service revenues. As in the inverse hockey stick theory of network build-up, the cellular network is built up first for coverage, and then for capacity. Also, as in the hockey stick theory of service usage, the service usage – and its revenue – starts off at a growth line, and hits an inflection point after which the usage rate increases quite significantly. The Double Hockey Sticks theory also illustrates that the starting point of the two curves is not the same – service usage will not start at the same point when the first network element is deployed, but rather several months after that – when an adequate coverage has been achieved in the network rollout.

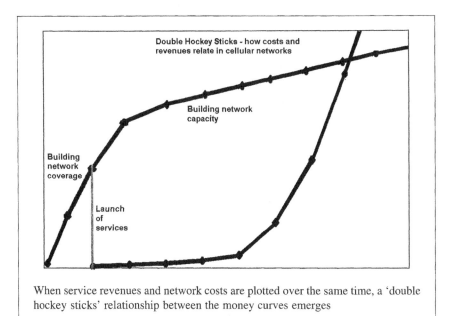

When service revenues and network costs are plotted over the same time, a 'double hockey sticks' relationship between the money curves emerges

The initial network costs seem excessive until the build-up inflection point is reached. Similarly the early revenue growth seems too low to sustain a profitable business until the mobile service revenue inflection point is reached, and the curve takes a significantly stronger growth pattern.

11.6.1 Uncertainty period

The early data points on the cost and revenue projections create preliminary data which are not favourable to validating the business assumptions. This creates a period of uncertainty, or 'Window of Worry'. The uncertainty period starts when enough of the network is built so that services can be launched. As cellular networks are always built first for coverage, and later for capacity, it means that any early costs data points will result in an over-estimate of the total costs of the network. And as cellular services follow the hockey stick curve, it means that any early service usage and revenue data points will result in an under-estimate of total usage and revenues.

It is very easy for those not familiar with cellular network building costs, and cellular network service patterns, to make hasty generalisations about the overall costs and revenues. Such analysts will have very valid early data to 'prove' their case. This type of early pessimism precedes most cellular networks' successes and was typical of GSM business cases in the early 1990s. Even as the management of the network operator tries to convince owners and bankers on their valid business case, such uninformed outside opinions easily provide for added anxiety in such cases. This is why the time is called the Uncertainty Period, or why there is a 'Window of Worry'. Of course, once the inflection points are passed, and especially after the service revenues start to increase dramatically, the concerns are finally removed. It is easy of course, at that point, for the early pessimists to say that the change in service usage was a surprise, that early data did not support such a dramatic turn in usage and revenues.

11.7 Borderline

Recognising significant and repeating patterns in a new industry helps in planning. That is why I wanted to include the Double Hockey Sticks theory in this book. It does give a solid theoretical framework for understanding the dynamics of the evolution in both the network build-up costs, and the result-ing service traffic and revenues. It should not take 3G network operators by surprise; and in most cases it will not, as they are familiar with these concepts. The service partners and application developers need to be aware, partly to be prepared to scale the service dramatically faster than the early take-up. And all parties should also allow for the hockey sticks when planning revenue sharing contracts and trigger points.

This chapter has also examined the new phenomenon of Reachability. Understanding the patterns is critical to achieving plans which reflect the reality and can be met. Those who understand these concepts well can best capitalise on them. As one of the world's greatest experts in rapidly recognis-ing patterns and adapting himself to them, the hockey great Wayne Gretzky used to say: "I skate to where the puck is going to be, not where it has been."

12

'The skilled executive can negotiate in such a way that all the parties believe that they have won the largest slice of the cake.'

Lewis Mendelbaum

Tariffing:

Just below the pain threshold

This chapter looks at tariffing for mobile services. Tariffing is the single biggest element determining the profitability of the whole business. Tariffing has a bigger impact than network dimensioning, sales provisions, handset subsidies, or any other decisions made relating to mobile services. It is worth repeating: Tariffing is the single biggest element determining the profitability of the whole business.

It may be the one area out of mobile services marketing that mobile operators and content providers *think* they know the best. Recent history shows that dramatic and even catastrophic tariffing errors have been made all over. Tariffing is often an afterthought, or left to a novice product manager fresh out of university, or a relic of history resting on concepts of some service idea from long ago. Tariffing directly affects the profitability of the company. That is why this chapter is needed. It also is the foundation to explain why segmentation brings profits. Before one can understand why segmentation brings profits, one has to understand telecoms tariffing, and concepts such as opportunity cost and pain threshold.

12.1 Some customers are willing to spend more

The underlying premise for tariffing and segmentation is that some customers are willing to spend more money for a similar, sometimes identical service than others. For a service provider in any industry, the ability to identify which customers are willing to pay more, and developing service categories, bundles, tariffs and other means to enable charging more for some is a key to profitability. Some early data are very promising on the mobile services opportunity as Deutsche Bank reported in its Wireless Internet report.

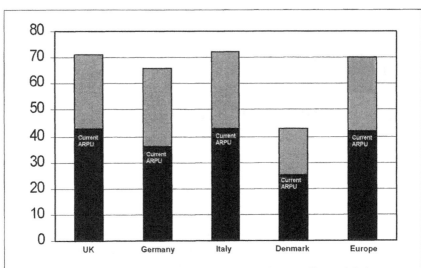

Consumers are willing to spend around 50% or more above and beyond their current spending, to purchase new wireless services, according to Deutsche Bank Wireless Internet report May 2001

The need to fill capacity is even more important in a capacity-driven industry such as telecommunications. The network is built, it has a certain capacity, whether nobody places any calls, or if all capacity is used. From the costs of running the network there are very few chances of variance in costs to operating the network.

12.1.1 Airline analogy

The mobile telecoms industry is thus very similar to the airline industry. If American Airlines has scheduled a flight from Los Angeles to Tokyo, it will most often have to fly the plane whether that particular plane is sold to 95% capacity or to 15% capacity. Odds are that there is also a return flight where there are seats sold from Tokyo back to Los Angeles, and even if one flight happens to be near-empty, odds are that its return flight is again more than half full.

The need therefore, is to fill the planes. And to try to get as high a price per passenger as possible. This is why there are several flying classes, such as first class, business class, improved tourist class, and tourist class. Actually, if you examine the prices paid per seat on any of the classes, there is considerable variation where one passenger might very well pay over double for an identical 'class' seat than another. The price will vary not only by class of seat (class of service), but also by time – usually if you buy very much before the flight, such as 3 weeks before, you get cheaper prices than if bought on the week of the flight. Some prices are promotions that have limited seat quotas. The price varies by location of purchase, typically the same Los Angeles–Tokyo round trip flight will have a different price if bought in the US or in Japan (or from elsewhere). The price varies by bundle – if you buy an American Airlines ticket from Miami via Los Angeles to Tokyo, the total cost will typically be less than if you bought one ticket from Miami to LA, and another ticket from LA to Tokyo.

The ticket price varies by distribution channel (different travel agencies may charge different prices) and purchase method (Internet purchases are often at a discount) sometimes by payment method (a credit card company may have a pricing special) or by customer affiliation (membership for example of a tourist association or citizens group may entitle you to a discount). Seats are sold also on bulk discounts (group fares, tourist agencies use this method to fly groups of travellers to popular destinations). Then there are the wholesale clearance resellers, who fill seats. These include all kinds of innovations such as bids and auctions. And then there are prices customised by customer. The airlines offer special prices and deals to their frequent fliers, offer occasional free upgrades (hence *de facto* discounts), and free trips after the passenger has flown a specified number of miles, etc.

The number of different tariffs for one route may seem mind-boggling. But if one person needs to fly to a given destination, and knows it 2 weeks before, and another needs to fly to the same destination, but only finds out 2 days before, it does make sense to 'punish' the late person by forcing that person to pay more. Odds are that most alternate and possibly cheaper seats are already sold. To protect the tariffs, airlines have invented numerous rules into their special prices, such as the Saturday overnight stay at the destination. The difference between a regular tourist ticket (which few passengers actually pay) and a discount tourist ticket (which most tourist class tickets tend to be) is that rule necessitating a Saturday stay. This balances loads away from the most frequent days of travel.

12.1.2 Applying example to telecoms

Can this principle be used in telecoms? Of course it can, and in small ways it already is done today. Let us imagine that you are in France and have a need to contact a friend visiting Brazil. First let us assume you are a German businessman and your employer pays for all of your mobile phone calling costs, including personal calls. You would have no problem calling your friend in Brazil. If your employer does not allow you to place personal international calls, then you would consider the cost of calling from the hotel room – and perhaps consult the hotel calling guide to see how expensive it is. Depending on your propensity to spend and the need of the contact you might call, or you might for example send an e-mail message to your friend rather than to call. But you probably would not bother to search for the lowest cost calling cards, i.e. you would essentially consider is the relatively expensive hotel room call 'worth the contact'. You probably would not consider renting a mobile phone to be able to place the call. You might consider going to the payphone and perhaps using your credit card or coins to pay for the call. The rates for coin calls from a phone booth would often be less than the price from the hotel, while the call might be handled by the same telecoms operator.

In the same market there are numerous operators which offer calling card services by which one can place much cheaper calls. If you were living in France, or visited it often, you probably would get such a calling card. Now we have several call prices for the same route. Actually there is a wide variety of other pricing elements, such as congestion – calls during office hours tend to be more expensive than during the evening, and discounts for contract customers, etc. In the above example we assumed the calling person and his condition is the main element for allowing change of tariff. Actually, the reason for the call may be a more significant element. If we change the picture just a little bit, in that the person in Brazil is your boss, and he is about to make decisions on your yearly bonus, and has asked you to call on this day, then price of the call will no longer be an issue. You would be willing even to rent a mobile phone just to be able to call your boss, rather than get less of a bonus than you feel you deserved.

The above example illustrates both that telecoms operators already use variety in pricing a similar or even same service, and that users will have different willingness to pay, and that willingness may even change per person depending on conditions.

12 Vignettes from a 3G Future: Perfect Radio Station

The 3G streaming music is fantastic. There are hundreds, or probably thousands of targeted music streams, and all are accessible with my mobile phone. Since I carry my mobile phone with me anyway, I no longer have to carry any other music player or radio, and I get better music than if I brought my favourite CDs with me. I tune into the station with exactly my music taste and mood for the moment, and can be sure that I get what I wanted. Sure, there are the occasional ads, but this is radio and it is free.

Music streaming is still seeking its value proposition. We already have free FM radio coverage and numerous ways to consume digital music. Music streaming brings several benefits to the music experience such as finding favourite music through the methods pioneered by Internet bookseller Amazon.com – other people who read this book also enjoyed books by these authors. The more the system learns of an individual's interests, the more it could tailor music approaching a segment of one – where a radio station would have to cater to its listener demographics. A further benefit in streaming would be the forced automation of listening to ads – you could not switch to another channel until the required number of ads have been heard, greatly increasing the effectiveness of advertising, reducing the need of repetition in ads in streaming.

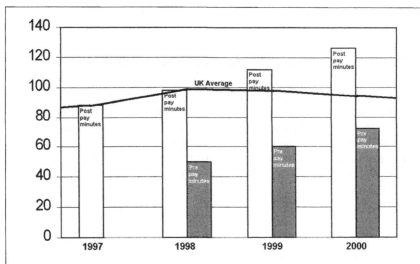

Minutes of Use (MOU) overall have remained relatively stable, with prepaid use and postpaid use both growing to compensate for the increasing proportion of prepaid accounts
Source: Deutsche Bank Wireless Internet Report May 2001

To understand who the customers are, what are their needs, how your services can address those needs, and what the customers are willing to pay for your services is vital for ensuring that you maximise your profit. Some customers in some situations are willing to pay much, and you should make sure that those who are willing to pay the most will always have easy access to your services. Other customers are not willing to pay as much, and you should have various flexible means to maximise your traffic and utilisation, but also to make sure that those paying less will not crowd out those willing to pay more.

12.2 Profit and pricing

When simplified, Profit equals Sales Price minus Costs. It may seem very simple to say that you can increase your profit by raising your price, or of course by lowering your costs. What is also comforting is to see that the laws price and demand seem to apply in a common-sense way, that roughly in proportion to prices falling, mobile telecoms usage has been growing. This was documented for example by Deutsche Bank in its Wireless Internet report when examining the UK mobile telecoms market and the advent of pre-paid customers. With the overall drop in prices, the usage kept going up keeping total revenues in proportion about the same.

While voice tariff can be expected to follow existing prices and adjust gradually, new services have no such past history and pricing has to be set 'in the dark'. The problems start with getting the initial pricing wrong, pricing too low, and then if forced to raise prices the customers would be driven to your competition. The risk of pricing too high is that nobody will use the service, and repeated lowering prices may be seen as an 'admission' that the service is no good anyway. With new mobile services the issues get even more complicated as we are not looking at only one service. Mobile operators and their service providers have to consider each service separately for tariff and profit.

Each service needs to be considered separately looking at its competition, and examining the price elasticity for that service. With pricing there are two costs. If the price is set too high, there is a cost of losing customers. If the price is set too low, there is an opportunity cost of lost profits. We have already seen from the airline pricing example that different consumers are willing to pay different amounts, and even the same consumers are willing to spend different amounts depending on the conditions. This concept needs to be applied across all services with price optimisation held in an ever-adapting competitive landscape.

The effects of optimised pricing are quite dramatic. A difference of 1 penny in pricing can have a huge effect on a popular service. For example if we assume a new service is priced at 14 cents per use, and it is rapidly adopted and soon used by 20% of the subscriber base, who use it once per day. If we assume for the sake of simplicity that there is the ability to charge 15 cents for that service at the same levels of use, then let us see how that affects a typical operator in a large European country like Italy, France and Britain. A typical network operator would have about 10 million subscribers, so 20% of that is 2 million users of the service. At one use per day, the service would be used 30 times per month or 60 million uses per month in the network. If the price was too low by 1 penny, the network operator – and any related content and application partners involved in a revenue share – would abandon 1 cent per user per day, or 600 000 dollars per month (60 million times 1 penny = 600 000 dollars), i.e. 7.2 million dollars per year. As we saw in the micro-payments chapter (Chapter 4), the tiny amounts, multiplied by the dramatic subscriber numbers, yield large revenues. This same magic of micro-payments works also in the profit and tariffing calculations. Every penny counts, very significantly, in mobile services.

Table 12.1 Typical service pricing – Iobox[a]

Service items	Price UKP	Price USD
E-mail forwarded to mobile	0.19	0.28
Send an SMS message	0.06	0.09
Calendar reminder via SMS	0.19	0.28
Send event by calendar	0.19	0.28
Send contact by SMS or card	0.19	0.28
Icon, tone or picture message	1.00	1.50
Mobile chat	0.06	0.09

[a] Source: BWCS Mobile Matrix (2002).

12.2.1 Pricing of bundles ('service packages')

A particular early favourite of new service portfolios is the 'service bundle' sometimes called 'service package'. Bundling services makes bundle pricing an issue. It is too easy to collect a dozen services, bundle them together, and then say "I'll give 10% off the whole set if you take the bundle." Telecoms operators have been experimenting with this type of bundling pricing with long distance and friends-and-family, etc. types of early service bundles/packages. Bundle pricing is its own special science, but the very short discussion is this: make sure every component in the bundle is actually needed in the bundle, and that each service is separately analysed for network load and profitability. A good bundle attracts customers, binds them to the operator, and allows use of *new services* that the subscriber otherwise would not trial or use. A bad bundle gives extra discounts to customers who would use those services anyway.

12.3 Preparedness for tariffing

Tariffing is critical to profitability, yet its margins of error are very small. Price too high, and you lose customers, or the service will not take off. Price too low, and you give up profits. How can the 3G network operator solve this dilemma? The solution is a mixture of market research, tariff modelling, trials, and adaptation.

12.3.1 Marketing research

The first step is marketing research. Any services which have a significant impact on the network should be measured through professional marketing research. Surveys and opinion polls can be conducted to test various service tariffing *types* – i.e. per usage price, monthly price, sponsored service, etc.; and especially to test the tariffing *levels*. Tariffs should be tested not only in direct questions such as 'how much would you be willing to pay for this service' – but also through indirect means, such as testing the price level against existing service prices, such as SMS (Short Message Service) text messaging, etc. For any existing services by competitors, the competitor prices are of course also of great importance (Table 12.1).

The network operators should build relationships with one or two major marketing research companies to build consistent surveys using the same procedures for the service portfolio. Results should be directly comparable across the service portfolio, and over time. major content providers should find marketing research partners that can offer services in many of the content provider's major target markets, to provide results which are comparable across international borders.

12.3.2 Tariff modelling

A very useful tool for evaluating tariff effects to profitability on a portfolio of telecoms services is Tariff Modelling. Many telecoms product management departments have rudimentary simple tariff modelling tools, mostly simple spreadsheets. More advanced Tariff Modelling tools are available from some of the specialist consultancies serving the telecoms industry. In designing a new portfolio of services, such as in introducing 3G services, the telecoms operator is prudent to use powerful tools to optimise its tariffing structure. Remember that a penny off from optimised pricing is easily millions on an annual basis.

12.3.3 Tariff trials

When introducing new services, it is crucial that tariffs are announced as introductory and that tariffs are adjusted as user data comes in. For the good-will especially of the 3G operator, which is perceived to be controlling the prices of all of the services that can be consumed on its service, it would be good if the 3G operator would have approximately as many prices going down as going up, after such trials. The operator should take the introductory pricing decisions as a learning process, and try to get the initial tariffs as close to right as possible, rather than every time giving it for free for 3 months and then introducing an astronomical price.

The time to use a 'try it for free for a month' or more approach, is when the service is genuinely new with no obvious near substitutes. Video calls on 3G networks could be such a service, as well as various multimedia messages. But if the service is only a minor adaptation of a similar and popular service on the fixed Internet, or in the newspapers, radio, etc., then there is no real reason to give it for free initially. The tariffing managers should be motivated to hit the right tariffs from the start, to invest good research, analysis and reasoning to find the optimal tariffs from the beginning.

12.3.4 Tariff adaptation

Finally tariffs will need to be adapted to the customer feedback and to market pressures. Here it is vital that the service creation system has a rapidly adopting structure to change prices. Many telecoms prices are so deeply ingrained into switching systems and monstrous charging and billing systems, that a price change literally takes months to implement. That is totally impossible in the 3G services area. Changes in tariffs have to be fast and nimble. Over times tariffs for homogeneous products tend to harmonise, as has been shown by the UK regulator OFTEL in its study of the average price of mobile call minutes in the UK market.

12.4 How about one price for all

In every major telecoms market with open competition there are players who are offering 'only one price' often suggesting it is both simple and cheap for the user. When one examines such operators, it usually emerges that the 'one price' is only offered for a mainstream mass-market product (such as long distance voice calls) but even these operators have special services priced differently. In most cases they are also willing to offer bulk discounts or other incentives for particular customers, such as large corporate customers, which means that the argument for one simple price is actually only a ploy to convince one part of the total customer base that this operator is the easiest to use.

One price for all can work for a player even in the mobile Internet services market if that operator offers a simple portfolio of mostly mass-market products, does not compete with high innovation services, and is best suited for a player aiming for price leadership. But we have to keep in mind that in the mobile Internet space even for a limited portfolio of services, there will be

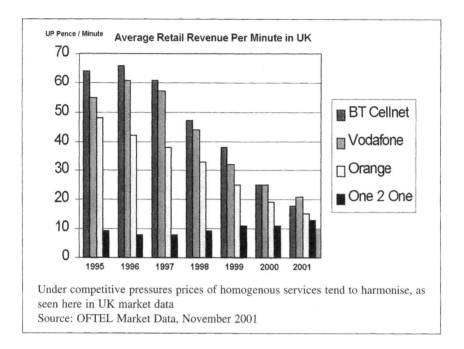

UP Pence / Minute   **Average Retail Revenue Per Minute in UK**

Legend: ■ BT Cellnet, ▣ Vodafone, □ Orange, ■ One 2 One

Under competitive pressures prices of homogenous services tend to harmonise, as seen here in UK market data
Source: OFTEL Market Data, November 2001

numerous ones where strictly speaking one price may be impossible. For example m-commerce, if we have a single set price for 'all' activity on the mobile Internet, what if I access a vending machine to purchase a Pepsi? Is there still a separate charge for the can of the beverage or will all such small transactions be bundled into my one fee. When mobile services were mostly voice and SMS traffic, the argument for a simple price had a lot of merit. But how about sending picture messages from my phone to a friend who is on another network? Listening to music which is streamed from some 'm-DJ' on a virtual m-radio station in New York City? The scope of different services which are all working on different value systems is so broad that a simple 'we have only one price' may be a nice thought but very difficult to achieve without a lot of customer confusion and complaints.

12.4.1 Home Zone

To compete with the local fixed line provider, the mobile operator could offer all calls made from home to be at the same rate as the fixed calls are with the competitor. This could be done with a feature called 'Home Zone' by which a designated area is identified as the home, and whenever the person uses the phone in that area, its tariffs would be the same as if using the fixed line phone. This might not be very profitable if the person never leaves the home, but the assumption with Home Zone is that the person will take the mobile phone also with him/her whenever going to town, etc., and at these times the mobile phone tariffs would be in effect. The profit calculations need to be carefully made to ensure that this service remains profitable in the longer run.

12.5 One way or another

In tariffing mobile services it is vital to keep in mind that mobile Internet services for the most part are mass-market services. Their true power is unveiled when hundreds of thousands, or preferably millions of people use them frequently. The pricing will need to be set very low not to hinder usage and to allow word-of-mouth and viral marketing to take care of publicising the new services. To summarise this chapter, remember the subtitle 'just below the pain threshold'. The clear example of how powerful this can be is SMS text messaging. As text messages typically cost between 10 and 15 cents per message, they are comfortably below what almost any person would stop to consider as a prohibitive cost per usage. Of course when the phone bill arrives and it has 50 dollars of SMS text messages, we may stop and think we should cut down, but at 10–15 cents per message, the individual cost is simply so small that we don't bother to worry about it.

That is how mobile services should be tariffed. Keep the price below the pain threshold. If not sure, aim *lower* and then build the traffic and adjust later. Don't stifle the usage by tariffing too high. A good example of how wrong operators got their pricing was WAP (Wireless Application Protocol) pricing. Here users were waiting to receive some information from a service which seemed extremely slow, and became increasingly frustrated and even angry when considering that every moment of delay was costing the users. While pricing is not the most glamorous part of the telecoms business, it is none the less the one with the biggest impact on the bottom line. While pricing managers get lumped in with accountants don't despair as comedians make them the objects of jokes, such as David Letterman who says, ''There's no business like show business, but there are several businesses like accounting.''

13

'People will buy anything that's one to a customer.'
Sinclair Lewis

Marketing 3G Services:

Segment, target, bundle

Modern marketing methods will be critical for success in 3G. Any 3G operator will be facing fierce competition, from more competitors than it is use to, in an untested technology, and in an environment which allows for totally new services. The competitors are increasingly globally backed, reflecting professional world-class competition. The 3G operator will need to recoup heavy investments in the infrastructure and in many countries also to recoup high licence fees. The 3G operator may have owners who may have a need to get a fast return on their investment. All of these factors place higher value on effective marketing of the services. To build a framework on which to consider the new marketing environment, let us start with an analogy from the retail industry, and look at the supermarket example.

13.1 From door-to-door salesman to supermarket

Until 1999 most mobile operators typically offered only about a dozen or so services. These included calls within their own network, calls to other networks, calls to international numbers, internationally roaming calls (calls made when travelling abroad), SMS (Short Message Service) text messaging, data access (modem access), and a few so-called advanced services such as VPN (Virtual Private Network) or Home Zone or 'Friends and Family' services. The total portfolio of services was very limited, a dozen or so services. The operator could spend a lot of time developing these few services, and dedicate specialists to manage each of the services, often having product managers for various segments for each of the products.

To draw an analogy from the retail world, this would be like the door-to-door salesman, selling a couple of different products, such as a small selection of cleaning chemicals, or a selection of brushes, etc.

The 3G network operator will be living in a totally different world. With the vast variety of services, service providers, content, content providers, and technologies, the operator will typically be launching hundreds of 3G services, and within 2 years managing thousands of services. By this it is important to notice that selling movie tickets is quite different from selling train tickets. The partners involved, their marketing and billing, and any mobile advertising, cross-marketing and m-commerce involved will mean that in both ticket sales there are different partners all along.

While a consumer may think that buying a ticket to see a movie is almost identical as a service to buy a train ticket, for the mobile operator all of the partners and participants in either value system will be different. Hence these two have to be considered different services. The same is true of tickets to sporting events, airline tickets, etc. etc. etc. That multiplied hundreds of times across information, entertainment, communications, commerce, etc., yields a portfolio of several thousand different services.

Here the analogy to retail is a supermarket where they sell not only similar brushes and cleaning liquids as the door-to-door salesman, but the supermarket sells probably numerous brands and types of cleaning liquids and brushes, as well as milk, bread, cheeses, etc. etc. etc. To understand how drastic the change is for the shopkeeper, imagine if the door-to-door sales- man all at once was asked to set up a supermarket from scratch. Not to step into an existing supermarket, but rather to try to use his knowledge of brushes and door-to-door sales of a dozen separate items, and adapt that to managing a store with thousands of different types of items to sell. And that he could not go to any existing supermarket to try to learn how they do it. This is the situation that current mobile operators face.

In the analogy it is quite likely that door-to-door salesman used to buy his brushes from one source, but at the supermarket he would have numerous wholesalers and suppliers offering competing products. As the brush sales- man, all his samples fit in his case. He might take all of the brushes out to show them to his customer. But in a supermarket totally different rules of marketing apply. Now advertisements in local newspaper are an important vehicle to attract business. Window price specials are another thing he would not have been exposed to as a brush salesman. And very critically, as most supermarket managers will tell you, the *placement* of items in the store is often critical to their success. Placing ice cream topping next to ice cream, or potato chips near the beer, etc., will result in significantly more sales not to mention placing the candy near to the cash registers. Placement becomes critical when managing a portfolio of thousands of products.

This analogy of service supermarket helps to identify some of the major issues that the 3G operator will have to face. It will also help illustrate why some of these keys to success may not be inherent to the mobile operators in the second generation networks, and also perhaps show good industries from where recruiting can take place to find competent know-how for success.

13.1.1 Placement/aisles at the supermarket

Supermarkets tend to have the products for sale organised along the aisles of shelves and freezers and storage bins. The customer is expected to navigate the aisles. In larger supermarkets the aisles may be dedicated to only a given type of product, such as a whole aisle only for soft drinks, another aisle only for morning cereals, etc.

One way to look at the 3G service operator's need to guide its customers to navigate its portfolio of hundreds or thousands of services, is to think in terms of placement not unlike aisles in the supermarket. The portal design could be thought of as aisles. First you select a major category (aisle) such as news or banking or games, then select the level of the shelf – from news select headlines or business or sports. And then go to the individual items and brands, e.g. from sports select basketball or tennis or swimming, etc.

When thought of in this way, the portal manager's ability to organise, prioritise, and group services becomes very important. The services should be grouped in a logical and easy-to-use manner. The user ideally should respond to the portal and its organisation by calling it 'intuitive'. But to achieve such ease of use will require a lot of trial and error.

13.1.2 Loss leaders

Another example directly from retail, is that when you sell thousands of items, you can afford to sell a few for below profit, as long as you have a limit to how many the customers can buy, and with the assumption that they will end up buying many other things anyway to compensate for the loss on the individual item. A good loss leader is a product which has very broad appeal and is needed frequently. Ideally the loss leader is a product which expires naturally, and with mobile services a natural extinction time could be set, such as this service can only be used during October, etc. As the mobile operator service portfolio grows from a dozen or two dozen services to a thousand services, the potential to use loss leaders will emerge. The mobile telephony industry has to make sure, however, that loss leaders are not used in cash cows and services critical to profits, such as voice minutes, SMS messages, etc.

Many lessons can be learned by telecoms from studying the retail industry. Particularly creative and successful managers and innovators from the retail industry make good additions to telecoms service creation and portfolio management staff to bring in the know-how rather than learning everything from scratch using trial and error. The industry on the whole can also benefit from comparisons with the retail industry, and some of the issues that confront the mobile operators can better be understood when considering operator behaviour through the above door-to-door salesman to supermarket owner analogy.

13.2 Segmentation

Segmentation sits at the heart of marketing. The aim of segmentation is to learn enough about customers to be able to target the offering, and extract the best possible price from any distinct group of customers. If you ask a mobile telecoms operator today whether they use segmentation, they are likely to respond that they do. If you then ask how many segments the operator has, a typical reply is three, four or five segments. If you assume that a typical mobile operator may have 10 million customers, and you divide them into five groups, it means that approximately 2 million customers are 'lumped together' and generalisations are made about their preferences, behaviour, and propensity to spend.

While it is technically true that a total divided into five groups gives 'more' accuracy than addressing all 10 million customers as one, it hardly meets with the potential of modern segmentation. As competition becomes more intense, segmentation becomes ever more important in determining which players are profitable and which are not.

13.3 Some customers are willing to spend more

In the Tariffing chapter (Chapter 12) we saw that some users are willing to spend more than others. The concept is often referred to as the '80:20' rule in that 20% of your customers bring in 80% of revenues. While mobile operators are reluctant to talk about revenues by type of customer some data have emerged and Deutsche Bank reported the real world example from a mobile operator in Canada in how revenues and number of customers relate.

The profit information is even more closely guarded than revenues but the overall principles are the same. A small selection of customers bring in the majority of the profits as well. For the mobile operators the most important goal is to seek to find out which customers are delivering, and can deliver best profits. Segmentation is now the tool by which to identify which customers

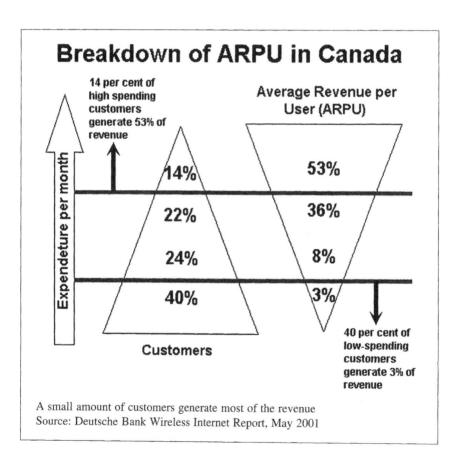

Breakdown of ARPU in Canada

14 per cent of high spending customers generate 53% of revenue

Average Revenue per User (ARPU)

Expendeture per month

14%   53%

22%   36%

24%   8%

40%   3%

Customers

40 per cent of low-spending customers generate 3% of revenue

A small amount of customers generate most of the revenue
Source: Deutsche Bank Wireless Internet Report, May 2001

are willing to pay what levels under what conditions. But before we can develop segmentation further, we need to start with examining existing segmentation.

13.4 Obsolete telecoms segmentation plans – size of customer

The classic way to segment customers in telecoms was business customers and residential customers. The business customers might be further divided by size for example into large corporate customers and SME (Small and Medium Enterprises) businesses. In modern segmentation models, this division by size is considered obsolete. The division into business and residential customers leaves an obvious overlap in cases where the business is a sole proprietorship, such as a carpenter, dentist, freelance journalist, artist, etc. And the division ignores similarities within parts of the two segments which can have similar needs across the segments but dissimilar needs within the segment, for example the early adopting technology-friendly users. The purchase decision and traffic patterns in telecoms between an early adopting wealthy private person, and an early adopting small business can be very similar to each other, and are probably very different from laggard adopter wealthy private person or laggard adopter small business.

> **Hint to operator marketing**. The traditional simple segmentation models are relics of history. If your company has one of the obsolete models, do anything you can to get the company to modernise. Your success is wholly dependent on it; your competitors are already doing it.

13.5 Segment by technology? Obsolete

Other often-used telecoms segmentation plans that are also obsolete, are those based on technology of the customer, i.e. fixed line customer vs mobile phone customer, or analogue cellular customer vs digital cellular customer, etc. There is a likelihood to extend this idea to 2G – 2.5G and 3G, or WAP (Wireless Application Protocol), GPRS (General Packet Radio System), HSCSD (High Speed Circuit Switched Data), etc. The needs of the customers are usually much more powerful motivators than their currently existing technology. So a travelling salesman's needs for telecoms services are probably similar to other travelling salesmen, regardless of whether one is using an analogue mobile phone and the other a digital mobile phone.

13.6 Segment by billing method? Obsolete

Yet another simple segmentation model used by mobile operators is the division into pre-pay and post-paid customers. Yes, there are many general differences between customers who pre-pay and those who are post-paid customers, but again the *needs* are more powerful determinants of behaviour than the coincidence of which payment method the customer happens to be using. If pre-paid and post-paid customers were a relevant *inherent* difference, it would not be possible to move customers from one group to the other, yet operators all around the world are involved in – and successful – in moving some customers from pre-paid accounts to post-pay.

13.7 Need to segment by need

The purpose of segmentation is to identify groups of customers whose interests – purchase decisions and spending patterns – are similar within the group, and distinct from at least some other groups. Segmentation allows for tailoring marketing and sales efforts to a group, addressing their needs without using reasons which may be meaningless for that group, or even worse – be against their interests. Without strong segmentation tools the mobile operator and its content partners are destined to fail in the competitive marketplace, in delivering desirable services, in tariffing for revenues, and in delivering profits. Modern segmentation is the key to targeted tariffing, and thus the key to success in 3G.

13.8 Example segmentation model

There is no correct answer to how many segments a model should have. The utility of adding further subsegments depends on the characteristics of the potential target market, and since we are all humans and our needs and wants change over time, the target market will also evolve. Segmentation models should have the ability to evolve as well. Segmentation model depth and sophistication is also dependent on the intellectual abilities of the various managers expected to use the model, so it cannot be too complex as to become unusable. When looking at a customer base of national size, easily in the many millions, the segmentation model cannot be too simple, dividing the groups into segments with populations still in the millions.

For the purposes of example for this chapter, a moderately sophisticated model is developed. This is not intended as the best model, only to illustrate

13 Vignettes from a 3G Future: Talking Maps

I could not imagine how useful the talking feature is on a map. My 3G phone knows where I am, shows me the map, tells me where to go, and if I start to go in the wrong direction, it can tell me I turned the wrong way. The map itself of course moves with me and shows my location in the middle of the map. I especially like the things that the system points out which are not easily visible on the map, such as landmarks, stores, gasoline station signs, etc.

The location accuracy in cellular networks is getting better and the features and utility of a location-aware service will depend on how precisely the phone can be pinpointed. Whenever a user requests a map, the system should also offer a click-to-hear button to offer machine-spoken guidance to the intended destination. The added traffic from the machine will be within the network and provide little additional cost, but if well designed, the service can add considerable utility to the user. As with most 3G services, the mapping and guidance systems are logical targets for sponsorship, for example by fast food chains that may want to entice people to stop by at the nearest establishment for a snack.

that even this type of model offers considerably more utility to the marketing and sales of mobile operators than a 'pre-paid/post-paid/business' customers or similar simple model.

The example model is five-dimensional. Any single dimension could be used alone, but the model gains accuracy and power by combining dimensions. Note that the model is intended for the next few years when voice traffic still dominates cellular services. This model examines only the needs of customers for voice services and messaging, so it will need to be updated and enhanced as more data emerges on preferences on data service usage. The model assumes a good database of customer information, so this model is most suitable for current incumbent operators who have a lot of data on their existing mobile phone users. For a totally new operator, a different model would have to be developed, as there would be no way to place any given subscriber into any given category.

> **Hint to content providers**. One of your first questions to a potential network operator (or MVNO) partner should be: what is your segmentation model? This tells you how archaic their marketing is.

13.8.1 First dimension: ownership

The first dimension, and very likely the biggest differentiating factor for phone behaviour, is ownership, or more precisely the extent to which the phone user is personally responsible for costs of using the phone. The simple question is 'who pays the bill'. The four groups are 'Company phone', 'Own phone', 'Second phone' and 'Parents' phone'.

A company phone user is one whose employer pays for phone costs. This person is usually least concerned with price, and usually has high usage. The own phone user is the 'typical' adult private person mobile phone user and the usage is likely to be most 'average' of the four groups. The second phone owner illustrates peculiar usage differences from the other three groups, in that the second phone is used only at certain times, or only to receive calls, or for call-back behaviour (see below), etc. When the primary phone is also on the same network and on the same person's subscription, it is easy to find first and second phones. But mostly when people have two or more mobile phones, they are on different networks for price, coverage and other reasons. This can only be identified by studying the calling patterns and identifying what are distinctive forms of second phone behaviour. The last group in ownership is the parents' phone, for younger children. Here there are considerable limits to the amounts that can be spent.

13.8.2 Second dimension: geographic usage

A second dimension to the model is how the mobile phone is observed to move within the network coverage. These data are automatically collected for billing and interconnect purposes. Five distinct major groups emerge: 'Jetsetter', 'Driver', 'Mover', 'Commuter' and 'Homer'.

The jetsetter flies to other countries and uses the mobile phone in those countries. While only a very small portion of the total worldwide population, the jetsetter type can put a lot of high cost traffic on mobile networks, placing calls at airport lounges, from taxicabs, and while on business about the town in other countries. The phone bill for a jetsetter can easily be 5–10 times above the average phone bill. These users have particular needs and usually do not care much about costs, and are particularly concerned about mobile service availability when travelling. In many cases the jetsetter's efficiency while on the road is totally dependent on the mobile phone.

The driver is someone who sits in a car, truck or bus (or other vehicle) most of the day. This includes taxi and limousine drivers, truck drivers and any assistants that may be in the truck with the driver, bus, tram, train, subway, etc. operators and conductors, etc. The person characteristically places and receives calls very much while moving at driving speeds. As the telecoms traffic is from a vehicle cabin, typically there is less chance for data communication and most will be voice communication. If the work uses the mobile phone (for example to route a delivery truck along the day) then the mobile phone is indispensable. But if the work does not use the mobile phone (for example bus driver on a regular route) then the phone is a device to connect with friends.

The next type of geographic category is the mover. The mover does not mean to change residences, but rather of movement throughout the day. This could be a sales representative, or an expert working at different office locations during the day and the week, or for example a journalist, etc. The person places mobile phone calls (and often data traffic by accessing the Internet and/or company intranet) from many locations during the day, but is often in one location for longer periods of time such as 30 min or a few hours at a time. In this way the person differs from the drivers who are on the road all the time, and this type of worker can put a lot traffic on the network, especially if transmitting data to and from a laptop computer or PDA (Personal Digital Assistant), etc. This type of person is very strongly dependent on the mobile phone.

The next category is the commuter. Commuter here is not limited to using public transportation, a commuter is defined here as a typical worker who goes to the same work location every workday morning and returns every evening. The commuter can walk, bicycle or drive a car, as well as using public transportation. For the network operator and from the service usage point of view, the commuter uses the phone primarily in two locations: work and home. The movement is mostly between these two locations. Of course the user will have hobbies and interests and needs to go shopping, etc., so there is also movement in that sense, but this tends to be related to the after work return trip home, or in the evening after work. Note that students fit this pattern as well, and from a geographical usage point of view, a student commuter is no different from a worker-commuter. This type of person typically could handle much of the daily communication needs by fixed connections at work and at home. The mobile phone is more of a free-time utility device.

The last geographical category is the homer. A homer is a person who spends most of the time at home and has no clear other 'office' or other work/study location for the day. The person might be a stay-at-home parent or a home-office worker, or unemployed, or for example a freelance news photographer, or remote IT worker. This person's mobile phone behaviour is very closely linked only to the home as a base. This person might survive very easily with a fixed phone and may well do much of the daytime phone traffic through the fixed line at home.

13.8.3 Third dimension: calling pattern

A third dimension is calling pattern. This dimension has five categories: 'Sponsor', 'Initiator', 'Receiver', 'Avoider' and 'Call-Backer'. The sponsor is someone who takes other people's communication costs on to his phone bill. This could be the parent with the child's phone, where the parent says ''I'll call you right back so it comes on my phone bill'' or it could be the friend or spouse who has the employer pay for the phone bill. ''Let me call you and we'll put this on my company bill.'' The sponsor is typically a user who has lots of calls, and longer calls than average. The initiator is one who places a lot of calls and may also be receiving many, the receiver is one who does not initiate many calls, but gets proportionately a lot more coming in. Initiating and receiving are somewhat related to human psychology, some people like to take an active role in contacting friends and colleagues by phone, while others do not for various reasons. The initiating and receiving split also may be dependent on the conditions of work, where some are focused intensely and concentrate on longer projects, others are stuck in perpetual meetings, etc., while others are by role contacting frequently such as secretaries making arrangements for their bosses, etc.

The avoider is a person who is not obvious by the pattern of actual voice call traffic, but rather by other telecoms traffic clues. The avoider is the type of person who never answers the phone, is hopelessly busy or otherwise forever beyond any reasonable reach. Unreachable. That type of person has the voice mail always overflowing, the same people call the person many times per day until they get through, and – here is the obvious clue – people send SMS text messages to get the person to call. The avoider is an existing customer deeply in need of assistance, and any newer innovations in helping organise the telecoms traffic (and other needs) will assist this customer, starting with improvements to voice mail, calling groups, distinctive ringing, etc. The person is a ripe candidate for selling another subscription (for the important people), etc.

The last group is the call-backer. Call back in this case works like this. I am very low on funds and want to save on my phone bill. I call you for only a few seconds, we agree that you call me back, and I hang up. In a moment your phone calls me and we have a conversation for many minutes. The call is placed on your phone bill, not mine. Call-backers are people with calling needs but few funds. They can be children, students, unemployed or low-salary workers or those perennially in financial trouble. Call-backers are obvious candidates for services such as sponsored calls, ad-pay, etc. and various discount calling plans.

13.8.4 Fourth dimension: timing of calls

The next dimension relates to when the subscriber uses the phone. There are four distinct groups which can be easily detected from their charging records, 'All timer', 'Day caller', 'Evening caller' and 'Expat returner'. The 'All timer' places calls all during the day and evening and thus very likely uses that phone subscription as the primary or only one. The 'Day caller' calls only during office hours and almost never in evenings and over weekends. This person very likely uses the day phone for work, and has another mobile phone for personal use in the evenings and weekends. If this person pays his/her phone bill, the person is a strong candidate for specials on residential/recreational services to churn away from the competition who now get the evening and weekend calls. If the person does not pay for the daytime calls, then it is an employer's phone with restrictions on personal calls. Then any expansion of the service to cover leisure time should go through the company which pays the bills.

The 'Evening caller' calls only during the evenings and weekends and has very likely another work-related mobile phone. It is very unlikely that this person would add the work phone to the evening subscription unless the person is self-employed and becomes unhappy with the operator providing the daytime service. The 'Expat returner' is silent for weeks or months, and suddenly, mostly centred around a weekend, places a lot of calls day and evening. This person lives abroad but maintains a mobile phone in his/her home country, and whenever returns home, uses that phone a lot. This person is a natural target for services to roaming users and is probably very heavily committed to the network provider to keep the number, etc.

13.8.5 Fifth dimension: range of contacts

With the range of contacts there are primarily three types of callers: 'Few contacts', 'Many contacts' and 'Random contacts'. Contacts can be calls made from the phone as well as calls received at the phone. Few contact people are likely not using the phone at work, and have only a few people with whom to interact. These tend to be young children or very old people, or very isolated working age people. Few contact people have few people to inform, so they can rather easily change phone numbers and network operators, but if young children or senior citizens, they are probably not eager to bother to change operators. Few contacts could also signal a second phone, used as a private connection given to only a few of the nearest and dearest.

Example five-dimensional segmentation model
Ownership: 'Company phone', 'Own phone', 'Second phone', 'Parents' phone'
Geographic: 'Jetsetter', 'Driver', 'Mover', 'Commuter', 'Homer'
Pattern: 'Sponsor', 'Initiator', 'Receiver', 'Avoider', 'Call-Backer'
Timing: 'All timer', 'Day caller', 'Evening caller', 'Expat returner'
Contacts: 'Few contacts', 'Many contacts', 'Random contacts'

Many contacts people are somewhat typical mobile phone users, having many but mostly the same contacts. They differ from the Random contacts people who seem to call continuously new and different parties – and/or have continuously different people calling in. This pattern is most typical in all kinds of customer service jobs. It is very likely a work phone, and the random calls relate to work. There may be a set of common numbers which are continuously called, such as family members and a few work colleagues and bosses, etc., but the majority of the numbers called would only have a few calls at most per number, changing continuously. Any person who has a large number of contacts, whether familiar or unfamiliar, will be dependent on the phone number. This type of person will not need a large phone subsidy to upgrade, for example, as they are likely to remain with the same operator anyway because of the number of contacts who would need to be informed of a new number.

13.8.6 Usage of the model

First it is important to remember that any five of the above dimensions can be used alone, to divide the total subscriber base into a few groups. The model gains power when dimensions are combined. So for example if a person has a company phone, is a jetsetter, avoids calls, is a day caller and has many contacts, very much can be identified from this person, who is quite different from the person using their own phone, from home, is an avoider and evening caller with few contacts. In this way the model gains power by combining the dimensions and isolating individual user profiles. This model has a theoretical number of 1200 separate combinations, and the average segment size for an operator with about 12 million subscribers would be 10 000 users per segment. Of course some resulting segments are likely to be unpopulated (for example a likely very tiny or non-existent segment combination could be: parent phone + jetsetter + call-backer + expat returner + random caller) and others to still be very large, with up to half a million users or more (e.g. own phone + commuter + receiver + all timer + many contacts). Again, this model is shown here only as the starting point for development of more relevant and useful models that can truly serve mobile service marketing.

The model should be used in 'layers' relevant to any given part of the organisation. For sales the relevant segment dimensions and the depth of the model may be quite different from product development or from advertising or top management. Segmentation managers should have the model tuned to various audiences and bring greater depth or broader scope depending on audience. In the above example, removing three dimensions would scale the model to 20 segments which would be suitable for most general management issues, presentations, etc. Here the individual segment sizes would be about 600 000 subscribers.

The two-dimensional simplification of the model would not provide great insight to product development or sales. Thus when examining a given segment further, depth could be provided by further subdivision of the segment, and various service sales arguments and customer need definitions should be defined at the deepest available levels. For critical segments identified from the 1200 in the full model, dozens of critical segments could be examined much further, giving sharp understanding of the most important groups, perhaps of 50 000–100 000 of the most profitable, most visible, most spending, most loyal, most beneficial and earliest adopting customers.

13.8.7 Why segment?

Segmentation strikes at the very heart of marketing. All marketing activities – defined very broadly to include all product development, sales, etc. – should be considered from the perspective of the segmentation model. The model should be used to identify potential targets of marketing activities and help focus marketing to be most effective. The model can be used to increase revenues; increase profits; increase usage: minutes, sessions, megabytes, etc.; drive users to try new data services; upgrade subscribers; move traffic to more preferable services such as away from capacity-congested ones or to more profitable ones; reduce churn; get customers more committed; to get rid of unwanted customers, etc.

> **Hint to segmentation managers**. Don't ever tell your customers which segment they are, however much it may be of interest to you. Nobody likes to be 'grouped' so your customers will not want to know they are handled as a bunch of thousands, even millions of other customers. Keep segmentation discussions inside your company and its partners.

Segmentation allows the business to understand groups of customers much more deeply than just as one (or three or four) general group(s). Recently the evolution of marketing has brought about a lot of focus on segmentation and marketers like to try to isolate ever smaller and more accurately defined separate segments and clusters. And when doing so, marketers have been talking even of marketing to a 'segment of one'. Note that the above five-dimensional model would have typical model segment sizes still of about 10 000 users, and the variance between the smallest size and largest size segment would be considerable. Still this is much more precise than the traditional 'corporates, SMEs, and residentials' or similar rudimentary segmentation models with segment populations in the millions.

13.8.8 Multiple overlapping segmentation models

A model is a simplification of the real world. Thus no one model is perfect in every instance. The same is true of segmentation models. One should keep in mind that one segmentation model which serves a type of customer and proposition situation may work well there, but not be as good in another customer and proposition situation. There is nothing inherently wrong with having multiple overlaying segmentation models. By one such model I could be classified as an author – hence similar to a small office home office (SOHO) worker. Without meaning to put down writers and authors, the stereotype suggests that they are not early adopters of technology. But by another segmentation model I could be classified as a 3G business consultant. That type of person usually is exposed to a lot of the technology as it is being developed. By a third model I could be found to be a 'techno-geek' fan of emerging technologies and a probable target for experimenting with new technologies.

Depending on which model is used, better or worse understanding can be obtained on me. The segmentation experts need to create tools which allow for the multiple overlapping layers of customers' personalities and profiles, to help the marketing find the messages that really address the needs. If nothing else, at least the multiple segmentation models should flag the fact that this person fits several categorisations, and therefore it is possible he is not exactly like the stereotypical user defined by any one of them.

13.8.9 Advanced segmentation models

More modern tools exist which define microsegments on patterns which are not intuitively obvious, but can be identified by mathematical means. The limitations of traditional intuitive segmentation models – like the example built above – stem from the limitations of the segmentation managers in combining marketing, sociological and psychographic knowledge. The depth and utility of the segmentation model will be dependent on how well the segmentation managers understand the business and the customers, and how well they can define and describe these. Modern mathematical models ignore the labels and sociological 'reasons' for why people behave in a given way, they simply identify and isolate groupings of customers who behave in a similar way. It is up to marketers to give labels to emerging microsegments if needed. Such mathematical models are still in their early stages, but with subscriber numbers in the tens of millions, and the ability to rapidly identify new microsegments for rapidly evolving services, mobile operators and their partners should invest in new mathematical segmentation systems to gain competitive advantage.

13.9 Bundling

Bundling is an area of marketing which is still not well utilised in telecoms. Bundling will likely become ever more important as there will be larger portfolios of services to choose from, and as there will be an explosion of players offering some of those services through alternate means. For the 3G network operator it will become increasingly important to understand how to create and manage bundles and keep them profitable, and to use bundling as a tool both to build new business and to keep customers. To understand which components customers want in a bundle, marketing research is needed. Preferences can be ranked, etc., and experimental bundles created.

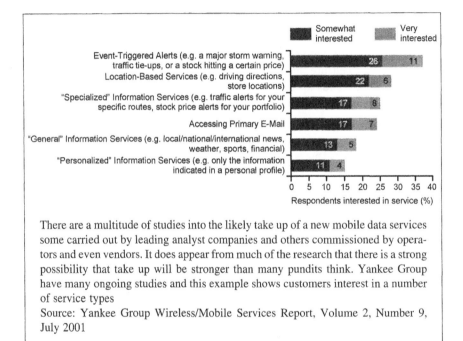

There are a multitude of studies into the likely take up of a new mobile data services some carried out by leading analyst companies and others commissioned by operators and even vendors. It does appear from much of the research that there is a strong possibility that take up will be stronger than many pundits think. Yankee Group have many ongoing studies and this example shows customers interest in a number of service types
Source: Yankee Group Wireless/Mobile Services Report, Volume 2, Number 9, July 2001

13.9.1 Bundling with (mobile) voice

The biggest single competitive advantage that the 3G network operator has over any players who do not offer mobile network services, is that of mobile voice (minutes). This should be used very carefully, in areas where the strongest bundling element is needed. The 3G network operator should be careful, however, to offer the same services also separately without the voice bundle, but at a significantly higher cost, to protect itself against competition regulators. A typical simple bundle example of a pure-mobile operator would be for a residential user to have home calls and access to the Internet in a bundle.

13.9.2 Home zone

To compete with the local fixed line provider, the mobile operator could offer all calls made from home to be at the same rate as the fixed calls are with the competitor. This could be done with a feature called 'Home Zone' by which a designated area is identified as the home, and whenever the person uses the phone in that area, its tariffs would be the same as if using the fixed line phone. This might not be very profitable if the person never leaves the home, but the assumption with Home Zone is that the person will take the mobile phone also with him/her whenever going to town, etc., and at these times the mobile phone tariffs would be in effect. The profit calculations need to be carefully made to ensure that this service remains profitable in the longer run even if the person ends up getting another phone from another operator, and its behaviour with your phone is skewed.

> **Hint to portal design.** The first thing to remember is that you are not a typical user. The typical user is an average person who does not know the difference between WAP and 3G, and will not bother to read manuals. Test your portal on your mother.

13.9.3 Portal power

Another strong advantage that the 3G network operator has in bundling is that of its own mobile portal. Practically all mobile operators have either introduced their own mobile portals, or announced an interest in doing so. The mobile portal itself might not provide much revenues or value, but if a good bundle is built around the basic 3G network operator portal, most users will not bother to seek others. Early reviews of independent portals trying to enter the mobile services arena have proven their case to be very difficult, and if mobile operators invest well into their own portals, they will keep the independent portals as a fringe player in the overall mobile services marketplace. As long as the 3G mobile operator's portal is seen as good, it also is then a powerful bundling platform. Here the operator should build a mix of services that they want to 'trial' for eventual launch as independently priced services, with that of a core set of 'anchor' services that should bind the user as much as possible. This type of anchor services could include e-mail, scheduler, SMS text messaging, some amount of multimedia messaging, etc.

13.10 Bundling with popular pages

Another way to use bundling is to 'ride the coat tails' of a successful service. Some content providers will probably bring in loyal customers and brand awareness. For example the sponsors of a Formula One racing team might find a strong association with the racing team, and use that in the 3G service side as well. For example Ferrari would carry a dedicated and passionate following, and probably Marlboro would like to continue their association with Ferrari from the racing car field also to the mobile services arena. Both Ferrari and Marlboro have Internet websites and will definitely also have mobile web pages. Cross-bundling works here as well, where at the Ferrari mobile website there could be a direct link to the Marlboro mobile website, and vice versa.

13.11 For your eyes only

This chapter has set the stage by examining the retail services model and applied lessons to 3G. The chapter looked at how 3G services should be marketed using standard theories of segmentation, deploying launch services with targeted and prioritised methods bring good returns. Operators must remember to cover all of the basics of their repertoire.

In deciding which services to create, the 3G operator should not try to create the perfect single service, but rather keep trying with lots of ideas that seem possible, to find the successful services. And one should not despair if initially not everybody sees the brilliance of a given idea. Many a fantastic invention was first dismissed by so-called experts as unfeasible. As George Bernard Shaw puts it: "The reasonable man adapts himself to the world. The unreasonable man persists in trying to adapt the world to himself. Therefore, all progress depends on the unreasonable man."

14

'Just when you think you've got the rat race licked – Boom! Faster rats.'

David Lee Roth

Competition in 3G Services:

More competitors

The mobile Internet space will see the traditional telecoms mobile operators facing more competitors with mobile licences, and numerous other competing offerings on other technologies and business models. One of the new players will be the MVNO (Mobile Virtual Network Operator) which does not own or operate a cellular network, but leases capacity from the network operators yet appears to the marketplace quite similar to the network operators. The competitive environment will be a difficult adjustment for established mobile operators, which in most markets have seen only two or three licensed carriers operating in the same market with hardly any MVNOs. This chapter looks at how the competitive market will become more diverse, and discusses some of the most likely strategies for the major players.

14.1 Types of players in the competitive environment

There are at least seven major groupings of players in the mobile services marketplace. They have different strengths and weaknesses, and all view the emergence of 3G as an opportunity to expand their role in the value chain. All players also feel the threats of other players moving into their traditional

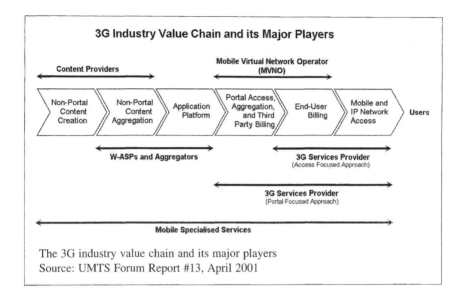

The 3G industry value chain and its major players
Source: UMTS Forum Report #13, April 2001

roles. One graphical illustration of some of the major players in the 3G value chain was offered by the UMTS Forum in its Report #13.

The mobile operator can be identified immediately from its licence. There will be only a limited number of mobile operators, mostly between three and six in any given market, and most – although not necessarily all – of these will have 3G licences. The network operator and the equipment vendor are the only players definitely needed in the system, although with 3G, the role of the content provider will become very important as well.

The second group is the MVNO, which may seem identical to the mobile operator when considered by end-users and customers. The MVNOs will not have a network licence but will offer mobile phone subscriptions and mobile services.

The third group is the content providers. The content providers could be players such as newsmedia, game developers, entertainment companies, music companies, etc. They have little interest typically in owning much telecoms infrastructure, but view the mobile Internet as one exciting new delivery channel. Content owners can have overpoweringly strong brand awareness and customer loyalty. A good example would be Disney, which already has numerous delivery channels and for its branded cartoon characters, no viable substitutes exist. It is up to Disney to decide if and when and how they want to participate in the 3G opportunity.

The fourth group is the fixed Internet ISPs (Internet Service Providers) like Yahoo and AOL. These will look to expanding their presence to the mobile Internet space. The mobile operators' own portals will try to keep the independent portals and ISPs in as small a role as possible.

The fifth group is the application developers, such as SAP and Oracle, who develop the information technology to enable services, and are seeking to gain from shifts in value chains. With vertical markets the role of the application developer can be crucial. As the 3G system gains maturity, some applications will be built into the system and the roles of some application developers will diminish.

A sixth group is the equipment manufacturers, which are building ever more features and abilities into the networks and terminals to take a bigger piece of the overall pie. Equipment manufacturers are incorporating features into handsets and the network platform and are even participating in content development. Together with the network operators, the equipment vendors are the only other players always present in the value system.

The seventh group is the various competing technical delivery means, such as digital TV, broadband Internet, satellite TV, and W-LAN (Wireless Local Area Network) also known as 802.11 or WiFi, etc. These will not be able to deliver exactly the same offering as 3G, but they can offer partial solutions which may be compelling, even superior to the 3G version of the same. There are also several other groups, such as portals, ASPs (Application Service Providers), web hosts, etc. But these could also be seen as parts of the above seven major groupings, so this chapter will limit the discussion to the seven groups.

14.2 Natural strengths

Each of the seven has natural strengths and parts of the value system where they are strong. These form the starting point to their competitive advantages, both absolute competitive advantages and comparative competitive advantages, as may be.

14.2.1 Network operator strengths

The network operator owns a licence to provide mobile service, and builds, owns and operates the mobile network. As such the network operator has a remarkable number of advantages. The first is limited direct competition. Every licence holder is known and there are likely to be very few in any

14 Vignettes from a 3G Future: Remembering the Face

I am terrible at placing names with faces. But I remember faces. Now finally there is a service that really appeals to me, it is the company 'face calendar'. Whenever someone calls me from work, I see the ID picture of the caller and the caller's name as my phone starts to ring. It really helps me with those callers who have only occasional need to contact me. And even better, if I am not sure which person any given name is, I can call up the face gallery on my mobile phone to view the face. It is perfect in the annual meeting of our associates from abroad.

The face gallery solution would be built around company scanned images, and could be combined with security badge ID photos. The images would be stored centrally and displayed through a standard called SIP (Session Initiation Protocol). The actual bandwidth required by transmitting a small 'thumbnail' image would not be great but the utility in any larger organisation would be considerable. The operator could expand the face gallery idea so that individual consumers could post their faces on to a central server. For the operator, showing a still image of the person calling shows vast improvement over 2G technologies and serves as visible proof of advancement in technology. On a more subtle side, showing a still image of the caller is of course a step towards accepting a moving image of the person talking, building acceptance of common use of video calls.

given market. This is mostly due to the limited spectrum available for mobile communications. The mobile operator controls the access to the mobile network. It is a natural gatekeeper position to control the services offered to subscribers, and to collect usage fees through the network operator's billing. A very powerful advantage is knowledge of the user through the charging information that is collected by the network. These are all absolute competitive advantages when comparing the network operator with the other types of players in the 3G services area.

Network operators also have numerous other benefits, often a well-established and trusted brand, a lot of resources both in technical engineering and in financial support. Many network operators have set up units or elements to take advantage of shifts in the value systems, such as setting up portals, banking services, advertising agencies, etc. In the long run a distinction between 3G and 2.5G network operators will become significant, when data communication capacity becomes a competitive issue. For the first few years where GPRS (General Packet Radio System) and EDGE (Enhanced Data Rates for Global Evolution) technologies can deliver a near-identical end-user experience as 3G networks, there is no meaningful difference between the network generations of operators.

Network operators will most probably be seen as potential competitors to all of the other six groups in this analysis, and in most cases network operators will also work in partnership with members from all six of the other groups. The network operator has a need of a lot of growth in understanding how to build lasting partnerships in such situations, and also to be prepared to see some of today's best partners becoming fierce competitors in the future.

14.2.2 MVNO strengths

MVNOs seem very similar to the network operator in the eyes of the consumer, but the MVNO does not own or operate a network. The MVNO will lease capacity on another network or networks, and most of the competitive advantages of the network operator are deprived from the MVNO. The MVNO can achieve some of the same advantages by negotiating with the network operator, but in such cases it will still be at the mercy of what the network operator will agree to allow. MVNOs can build their own charging and billing systems so some of the benefits can be theirs as well. MVNOs tend to be companies which are either strong in marketing with a strong national consumer brand such as the Virgin MVNO, or else a player with a leadership position in its own specialty such as the *Financial Times* MVNO. The

MVNO's primary competitive advantage stems from its relationship with its dedicated customers, and is usually very strongly correlated with the strength of its brand with its target customers.

The MVNO is seen most of all as a competitor to the network operator and of the six groups, the MVNO is perhaps the most obvious and strongest short-term competitor to the network operator. Depending on the MVNO's background, it may be in direct competition with a part of content providers or part of application developers, etc. They are likely to not view ISPs and equipment manufacturers as competitors, and if they are involved in alternate technical delivery systems, that too would most likely be in partnership mode.

14.2.3 Content provider strengths

Content providers hold the cards in allowing their own content on to mobile networks. The more the content is broadly available on other mass media, the better it is for the content provider. A typically strong content provider would be Disney, using a very wide array of media – conventional and non-conventional – ranging from movies and TV to comic books and newspaper serial content to merchandising and Disney stores, to Disney theme parks. But the more the content is replicated with homogeneous content, the less the content provider can bargain with its partners. A good example would be a provider of translations of Latin phrases. There is no copyright protection to the ancient Latin language, and any student of the language could set up a service to provide Latin translations. There would be little to differentiate one from another. It should be noted that many content providers will be considering the opportunity to set themselves up as MVNOs, like the *Financial Times* did.

Content providers are mostly competitors to the network operators and MVNOs, but can also have some competition with equipment manufacturers and application developers. Content providers are probably most openly willing to look into alternate technical solutions and can see them as a powerful ally in getting larger slices out of the digitally converged service delivery pie.

14.2.4 Fixed Internet company strengths

Fixed Internet companies, like Yahoo and AOL and thousands of smaller ones, are looking for revenues from the mobile Internet, especially after the Cybird example emerged in Japan. Fixed Internet companies bring experience in digital convergence, the e-business experience, Internet advertising know-how, and generally lightning-fast service creation and competitive reaction times. They will be facing well-entrenched telecoms operator ISP brands and companies and may find themselves with relatively little to offer as benefits in the mobile Internet marketplace. Some of the biggest ISPs offer impressive user numbers, but mobile operators have a much stronger connection with their subscribers than ISPs have with their customers. Fixed Internet operators would most likely set themselves up as independent portals or preferred second portals to mobile operators' own portals. The fixed Internet portals will be bringing in the majority of the portal revenues in the next few years, with mobile portal and digital TV revenues becoming significant over the later years, as Ovum predicted in December 2001 in its Future of Portals study.

Fixed Internet companies are natural competitors to (other) portals, both those of network operators, and any which try to establish themselves as independent portals. As such fixed Internet companies are rivals to network operators and MVNOs. Their technical background makes them somewhat able to compete also against application developers, and to a lesser degree against equipment manufacturers and content providers. Fixed Internet companies are naturals to use broadband Internet technology and may well be major players with other emerging technologies, especially W-LAN.

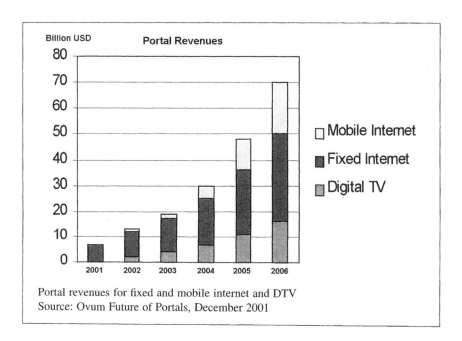

Portal revenues for fixed and mobile internet and DTV
Source: Ovum Future of Portals, December 2001

14.2.5 Application developer strengths

Application developers are working on ways to make their application smarter, and to retain more of the revenues relating to their technologies. As the 3G terminals grow ever smarter, some services which initially are set up as network services, can become more practical as stand-alone applications on 3G terminals. In this way application developers pose a long-term threat to network-centric services. In many cases application developers may set up their own niche MVNOs to take a step into the area of the network operator business.

Application developers are perhaps most competitors of the equipment manufacturers. They will have desires to take some slices from the network operators as well, but are rather small threats to content providers, fixed Internet companies and MVNOs. Application developers would probably view all emerging new technologies as opportunities to be exploited and thus strongly align with some emerging new technologies.

14.2.6 Equipment manufacturer strengths

The equipment manufacturers are looking to build more value-add into their network components and to their handsets and terminals. From a technical point of view, there is little to prevent an equipment manufacturer from building any of the services either into the network platform or into the handset as standard features. The biggest hindering factors are issues of global production scale – most country markets are simply too small. Nevertheless, equipment manufacturers have cautiously been moving in that direction already, and they pose a long-term threat, especially to any network-centric services which have particularly strong profitability. Equipment manufacturers have to be very careful, as in most cases the network operators are both their biggest customers for the network infrastructure, and their biggest distributors for the handsets.

Equipment manufacturers would be strong competitors to application developers, and lesser rivals to network operators, Internet companies and content providers. Equipment manufacturers would probably not see much competition from MVNOs and they would almost certainly offer some solutions in alternate technologies in addition to mobile Internet technologies.

14.3 Competing technical delivery solutions

Competing technical delivery solutions, from digital TV to satellite-based telecoms, to W-LAN (also known as 802.11 or WiFi), broadband Internet, Bluetooth, etc., pose the greatest uncertainty to 3G competition. Calculations on current assumptions of data throughput of the various technologies, of usage amounts and patterns, and a myriad of other elements, have been provided to argue on behalf of, and against, each of the alternate technologies. It will be impossible to say beforehand, of course, and only time will tell. It should be noted that unforeseen technological leaps and innovations can totally alter the service landscape of a given technology, with perhaps the Concorde/Boeing 747 being the most widely quoted example of how drastically a business proposition can be altered, turning the tables from the strong candidate in favour of the supposedly weaker candidate. It is very unlikely that any of the above-mentioned solutions would replace the 3G network and service portfolio as a whole, but any of the above could steal the most lucrative, or the most popular services and seriously dent the overall business of 3G. This chapter will take a 'quick and dirty' look at each of the mentioned technologies.

14.3.1 Digital TV

Digital TV has real potential to be mass market and as widely adopted as the mobile phone. Numerous content and commerce applications exist and are heavily promoted on digital TV. Digital TV will not offer voice or other easy person-to-person communication and of course digital TV will not offer mobility.

14.3.2 Satellite telecoms

Satellite-based telecoms suffer from bandwidth issues and terminal costs. The system was built to cover the whole globe, whereas most people do not live on the oceans covering 70% of the Earth, nor the deserts, mountains and jungles covering another two-thirds of the land mass. The solution has a hard time being competitive with 2G and 3G land-based mobile communications. But since the satellite systems are operational, their owners will try to find good use and traffic to the system.

14.3.3 Wireless LAN (802.11 or WiFi)

Wireless LAN, i.e. 802.11 or WiFi (Wireless Fidelity) has many good features to allow data communications within short to medium distances typically within a building. W-LAN is still evolving, and has numerous challenges relating to security, billing, roaming, etc. W-LAN is not optimised for voice and could have early difficulties in handling voice communications, but for high-capacity data transmissions W-LAN can easily outperform 3G in any given location. W-LAN will be developing fast partially because it operates in the unlicensed spectrum and also because W-LAN equipment costs a tiny fraction of 3G equipment, and thus many more manufacturers can afford to enter the game. Partly because of that, serious W-LAN solutions will have added problems of anybody else with any other innovation being able to design devices that operate – or cause interference – in the same unlicensed spectrum.

14.3.4 Broadband (fixed) Internet

Broadband fixed Internet has been evolving fast over the past several years and numerous companies have emerged – and died – in that space. Many have argued that it is relatively few of the new services that we really would want to consume while on the move and thus most of the so-called 3G services could just as well work on broadband and do not require mobility. If that is the case, broadband Internet could provide serious competitive pressure to 3G and W-LAN, by providing more bandwidth, lesser cost, and oftentimes more tried and mature services.

14.3.5 Bluetooth

Bluetooth finally, is the shortest distance wireless technology of those mentioned here. Bluetooth's range would be only to the nearest devices typically within a room. Bluetooth's appeal is in its projected very low cost and possibly near-universal adoption. Many of the types of 3G services that would work at extreme close range could be replaced by Bluetooth connection-enabled solutions.

14.4 Operator vs operator

As the mobile telecoms industry matures, its services will also become more and more similar. The need arises for individual players to find their place in the marketplace and focus on an area. Competitiveness becomes an ever more important factor in mobile telecoms. Traditionally it was not so, as still a few years ago all mobile operators were only involved in a hectic programme to get subscribers connected, and little or no real competition existed. In 3G there will be more competitors, with much less natural subscriber growth, and competitors will be more globally oriented, presenting a more world-class competitive challenge. In this environment operator-to-operator competition will become fierce, with a fight for customers, and managing churn becoming ever more important.

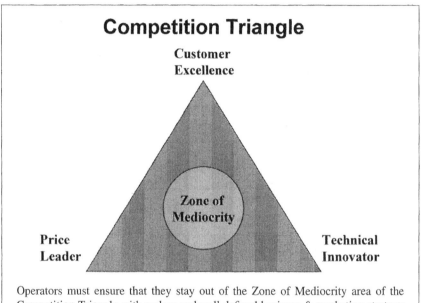

Competition Triangle

Customer
Excellence

Zone of
Mediocrity

Price
Leader

Technical
Innovator

Operators must ensure that they stay out of the Zone of Mediocrity area of the Competition Triangle with a clear and well defined business & marketing strategy

The competitive environment in 3G is likely to evolve so that there are a few distinct leadership positions of relatively healthy competitive position, and the players will be manoeuvring to get to the leadership positions. Typically this is thought to be a three-dimensional space, one is best quality (best network, best features, most innovative), another is best price (cost leadership, mass market), the third is best service (best sales and customer service). A player which attempts leadership in all three is bound to lose. A player which is average in all three will be most vulnerable. The player which can lock its dimension has the strongest chance to survive, and excellent planning and execution can even bring about leadership on two of the axes, at least for a short amount of time.

14.4.1 Price leadership

There usually exists a price leader or leaders in most markets. These have to have the most lean and efficient production capacity. They have to be a leading player in the mass-market area. They have large(st) volumes and can achieve cost savings through efficiency of scale. Price leaders tend to market through mass-market channels and they tend to feature their price leadership argumentation very strongly. They are vulnerable to new players who might offer lower prices, as lowest price is not a strong factor to build customer loyalty. Price leaders have to be efficient in cost and this usually results in modest facilities, lean and modern management structures and often a youthful organisation. One could say that price leader companies are also the favourite of controllers and accountants and tend to have a very strong internal influence exerted by the 'bean counters'. This rarely is the most profitable player but it can be the biggest player in the market.

14.4.2 Quality leadership

The leader in quality will focus on innovation, best on most quality measures such as radio coverage, dropped calls, etc., and often market heavily on their innovation and quality, possibly with references and testimonials. For the quality leader it is important to have modern platforms and modern development tools, as well as excellent R&D. Quality leadership takes time to achieve, but usually when achieved, it is relatively stable, as it is usually easy to notice if other players are attempting to break into that area of expertise. Quality leaders are the companies which are highly visible in the press with frequent new product launches and innovations, and tend to be the ones winning the quality and innovation awards. Quality leaders as employers are often the favourite player for new engineering graduates. Depending on the economic cycles this may be the most profitable dimension.

14.4.3 Customer service leadership

The third possible dimension to focus on is customer service leadership. It starts from excellent sales staff and processes, and follows on through excellent installation and customer service representatives, on to leadership in billing and reporting. As business customers tend to demand more service than residential customers, and as large, multinational and corporate customers are more demanding than business customers in general, usually the player which achieves customer service leadership is also the leader in the market niches of large corporations and multinationals. This company tends to be the favourite employer for business graduates. The service leadership position may be the most profitable dimension depending on economic cycles. This player usually has high customer loyalty.

14.4.4 Zone of Mediocrity

The worst position to be in is the one who is not the leader on any of the three axes. This player falls into the 'Zone of Mediocrity'. Players who are stuck here will feel the 'battle' on three fronts, and have to compensate for all. This is frustrating and consumes a lot of time and resources. Therefore, the players in the Zone of Mediocrity are almost always the least profitable of all. Players stuck in the Zone of Mediocrity may suffer from internal motivation and morale problems and suffer high employee turnover. Any operator stuck in the Zone of Mediocrity should set a clear target on one of the leadership axes and pursue that goal vigorously. It is the only way to get into a position of profitability. Also a word of caution, if this player targets the easiest of the paths – that towards price leadership – it may prove elusive as a goal and very short-lived as an achievement. The other two dimensions may be more viable as sustainable goals if achieved. The players stuck in the Zone of Mediocrity may be some of the biggest players, but their customer loyalty is usually very fragile.

14.4.5 Brand power

As services become ever more homogeneous, one of the last remaining possibilities for differentiation lies with the brand. Brand awareness has emerged as a major aspect of telecoms marketing during the 1990s and will become an ever more significant factor in the 3G market wars. As brand creation, development and expansion are all major areas discussed in many other books, suffice it to say here that remember to have a strong brand strategy for your service(s) also into the 3G space. 3G does provide the opportunity for new entrants to create the illusion that 3G is something dramatically different, where no existing brands have established themselves, and thus new players are on an equal footing. Equally any existing players who move into 3G will be able to bring their brand into play and expand it into 3G. 3G can also be used as the incident to introduce updates to the brand image for any existing players wishing to modify their brand perceptions in the marketplace.

14.5 Customer satisfaction and churn

A growingly important aspect of mobile telecoms marketing is customer (dis)satisfaction and churn. Churn means customers who leave one company to become customers of its competitor. Churn can be in your favour or against you, and mostly there is both churn out and churn in, and customer retention specialists talk of net churn as the difference of gained churners minus the lost churners. Churn has not been a significant factor in cellular telecoms in the past when the industry was experiencing hypergrowth. Now with competition showing more of the traditional features, churn is among the first to raise new alarm with mobile operators. The predicted amount of churning customers is quite dramatic. In Western Europe alone churning customers are expected to reach over 100 million annually.

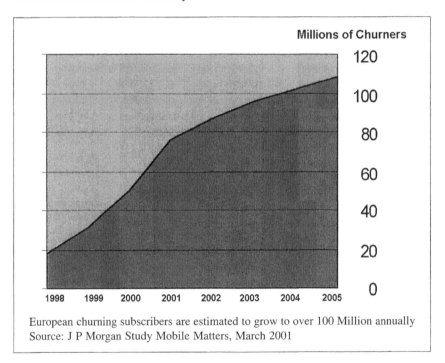

European churning subscribers are estimated to grow to over 100 Million annually
Source: J P Morgan Study Mobile Matters, March 2001

14.5.1 Know why

The first aspect of churn is to understand why customers churn. You might think you know ('the competitor ran a big discount campaign') but unless churn is tracked continuously and systematically, it is not possible to know for sure. Even then, marketing research, interviews, surveys, etc., have their limitations in accuracy when asking people why they stopped using a given company's services. It is very important to remember that you need to examine also the churn of your competitors. Only when you compare the customer opinion trends of your customers with the opinion trends of customers of your competitors, can you gain a solid understanding of how you are actually doing vs your competitors.

It always costs much more to gain a new customer than it is to retain an existing customer, so with the exception of some customers that you might not want to keep (e.g. the unprofitable ones) you should focus on keeping your own customers loyal to you.

Churn understanding is one area which in difficult economic times is easy to neglect. There is the temptation to think that since the data were collected in the past, we can skip some research now, and make up for it later. This thinking is most dangerous *in economic downturns*. Especially when there is a downturn, you will need to know *why* your customers churn – or don't churn – as well as why customers of your competition churn/don't churn. When overall resources are cut, it becomes ever more important to focus. Without knowing exactly *why* people make decisions on abandoning one operator and selecting another, you cannot implement effective actions to respond to that change in your existing and potential – and possibly departing – customers.

14.5.2 Sticky services

A way to keep customers locked into your service is having them signed up for what are called 'sticky services'. A sticky service is one which the customer will not want to change because of its effect on other parts of its business or interests. The classic example of a sticky service is the mobile phone number. Wherever number portability has not yet been introduced the stickiness of mobile phone numbers is still considerable. What makes it sticky, especially for corporate customers, is that the mobile phone number is stored and printed in numerous places such as business cards, stationery, phone directories, etc. If you switch services, you would usually have to abandon the phone number and take a new one. While the change itself might provide savings or convenience or better services, the change would also mean that you would have to reprint brochures, business cards, stationery, etc. etc. In addition to the direct costs of reprinting company literature, comes the opportunity cost of lost business. How many calls are lost because someone had the interest to call you, and had your card, but on the card was the wrong number.

Sticky services in 3G will be up to a lot of creativity and partly to luck. Some of the early promising candidates include the group scheduler and calendar, the 3G VPN (Virtual Private Network) service with its 'face gallery', etc., and the corporate portal. Any of these would include aspects which would be very tedious to go and change. But also the 3G and overall mobile Internet technology is evolving so fast, that soon these might be very easy to move, and the true sticky services could be others altogether.

14.5.3 Customer loyalty programmes – frequent caller plans

A special case of customer satisfaction is the emergence of customer loyalty schemes, similar to the frequent flier programmes of airlines. There is considerable novelty to innovations, so the real merits of a customer loyalty programme will be revealed with time. But in an industry where the cash cow products – mobile voice and SMS (Short Message Service) text messaging – are totally homogeneous, any way to differentiate your offering is better than nothing. Customer loyalty programmes can of course be bundled with those of other services in other industries, again the airlines have been pioneers in this with airline bonus miles collected from almost anything including hotel stays, car rentals, credit card purchases, even phone calls made on some calling cards. The ability to use the acquired miles is also expanding to cover many of the above.

But apart from 'earning' free minutes or services – or actual airline miles – there are other ways to bring benefits for frequent callers. One is the ability to 'cut in front of the line'. As there are few actual lines or queues where phone callers would physically stand in line, such as there are at airports where frequent fliers get lounge privileges and special counters to get fast service, at least the mobile operators can provide cutting in line features to their frequent callers to their call centres. So when a 'gold level' caller happens to call your calling centre, he/she should not have to wait and get the next available attendant. Benefits like these are also important in building the loyalty for airlines, and telecoms operators have a lot to learn from that industry.

14.6 Tariffing and price wars

Tariffing is a key element in competition, but less so in 3G than it was in 2G, as early on there will be a wide range of new services with which to differentiate. Tariffing was covered in its own chapter earlier on in this book

(Chapter 12). Perhaps a few words are needed on price wars. Don't start pricing wars unless you know how to end them and that you can win them. Price wars reduce everybody's profits. If you are in a pricing war, figure the way to get out of it as soon as you can. In 3G for the first 4 or 5 years there is so much new to gain that pricing wars are totally futile. Any effort used in trying to win pricing wars is just effort away from creating wholly new and profitable markets.

14.6.1 Profitability

Often considered a black magic science, shrouded in mystery and secrecy, the most important single attribute of a customer is of course profitability. It is astounding to find so many product managers in telecoms who openly confess that they do not know the profitability of their *own* service. Profitability should be a key determinant in selecting customers – and services. In an industry where the charging engine tracks the most miniscule expenditures, and where all equipment has to conform to rigid international standards, it is almost funny that profitability often seems to be an afterthought.

14.7 Go down fighting

This chapter looked at competition. It examined major groups of players and their natural strengths and their natural rivals and partners. The chapter also examined various rivals to mobile telecoms technology. It then focused specifically on how network operators can succeed in the marketplace and be profitable. Churn, price wars and profitability were also discussed.

Market success is mostly seen through market share, and market share should be one of the stated goals of any player in the mobile services area. But when seeking success in the marketplace, do not forget about profit. As Roger Enrico of Pepsico said: "Market share without profit is like breathing air without oxygen. It feels ok for a while, but in the end it kills you."

15

'Nothing knits man to man like frequent passage of cash from hand to hand.'
Walter Sickert

Revenue Sharing and Partnering:

When you cannot do it alone

The mobile services market is a drastic change from the former cellular carrier/mobile operator environment where the carrier/operator built and managed its own services from end to end. Mobile services will encompass initially hundreds and soon thousands of little services which in turn will form bundles and packages. It is impossible for a mobile operator to build all of these services, and it has to partner. For the content provider and application developer it presents an opportunity to build mobile services together with the big mobile operator on to its service creation/management/billing, etc. platforms.

In the chapter on the 5 M's (Chapter 5) it was discussed in detail how services can be made more appealing and profitable. This chapter looks not at how to make more money, but rather at how to divide up the resulting money. Here a preliminary word of caution. The mobile operator is the natural partner in most cases and on a basic level most executives at mobile operators do understand the need to partner. In the more practical day-to-day operations, the mobile operator is new to the challenges of managing multiple partnerships. The mobile operator will initially be cumbersome and easily make blunders in its dealings with potential partners. This is not due to malice, simply inexperience. The other parties who may be much more familiar with partnering in other industries and with other communication media should be tolerant of mobile operators as they climb the steep learning curve to become contributing and caring partners for the long run.

15.1 Know your dance partner

Mobile operators are not accustomed to working with dozens or potentially hundreds of small content providers and application developers. The content providers and application developers tend to approach the new service opportunity with perhaps an overoptimistic view of what can be done 'on a stand-alone PC' or other such simple service platform. These may be acceptable when users number in the hundreds or thousands, on a service which runs with an ISP (Internet Service Provider) for example. The requirements, however, of mobile operators, with easily 10 million subscribers are very different and amongst the first concerns of product development with major operators are reliability and scalability. Will the system stand a simultaneous request by 500 000 users? Will the new solution allow fraud into the network. Is it possible for the new system to crash the whole network. These may seem like paranoid concerns to a young game developer with the coolest idea on earth, but a cellular carrier cannot take the risk that its network goes down for even a few minutes. The traffic loss and bad publicity is not worth it.

On the other hand, mobile operators are not known for innovation and fast deployment of new services. They were used to product development cycles of 18 months, where their equipment vendors brought roadmaps of next evolutions of the technology 3 years into the future. That is a far cry from the Internet product development cycle, where 90 days is commonplace from idea to commercial launch. Here cellular carriers have to respond to market needs and speed up their development. Several partial solutions exist from specialised 'middleware' service creation platforms which allow fast implementation of solutions which then connect in a standard way to the various management systems, CRM (Customer Relationship Management) systems, billing and provisioning, etc. Another way is to have in-house test networks, for example in a given city or for a given customer segment, where services can be deployed on a smaller scale to allow trialling and learning.

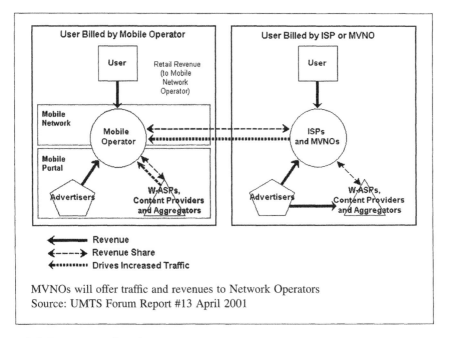

Revenue
Revenue Share
Drives Increased Traffic

MVNOs will offer traffic and revenues to Network Operators
Source: UMTS Forum Report #13 April 2001

15.2 Revenue sharing

The principle of revenue sharing is rather simple. Two or more partners bring in their own 'value-add' to the service proposition, and the partners share in the revenues from the resulting new business. When we examine this idea in more detail, numerous issues emerge. It is important to note that at the writing of this book, in the winter of 2001–2002, most players on both sides of the revenue sharing table were playing their cards very close to the chest, so very little information was in the public domain about exact revenue share ratios, etc. Therefore, much of the discussion that follows is from a theoretical viewpoint.

15.2.1 Start with definitions – what do we share

This may seem like a trivial point or one not worthy of discussion to those who have not been close to operators. The definition, however, of what will be divided is critical to understanding what money will be – and more importantly what will not be – divided among the partners.

Operators divide their potential revenue streams into revenue from transmission of traffic, and revenue from content. To explain in a few examples. If we sell a music CD single via a mobile phone, and it costs 5 dollars, then the operator might want to take a revenue share from acting as the payments processor for the sale, for example taking 5% as a sales commission. For the value of the CD the operator would be getting 25 cents. But the transmission of the data from the mobile phone to the record store would consume some seconds of on-air time, and some bytes of data traffic. This data traffic and airtime would be 'traffic' which the operator would not want to share. If the traffic was in peak-hour time, and included some 'surfing' at the record shop site, it could easily account for another 25 cents worth of airtime/data traffic. The traffic duration and cost are not predictable and can vary greatly. Note that the traffic portion can easily be more than the value of small-cost items, and that the traffic portion can vary greatly between different m-commerce shoppers. Some know what they want and go to it fast, others may linger for a long time comparing prices at various merchants before deciding what to buy.

The operator has a strong case for insisting that the traffic not be subject to revenue sharing. The operator has a licence to use the limited natural resource, the radio spectrum, and every bit of traffic that is placed on its network will eat up its resource. The operator can claim that mobile phone users will use their phones anyway for calls and data traffic, and that content providers must bring in an additional extra to qualify for revenue sharing. In the big picture the difference is more of semantics, but operators want to cling on to their sacred bit-pipe data stream revenues, guaranteeing the full cost plus profit on delivering traffic, even if all content migrated away from their control.

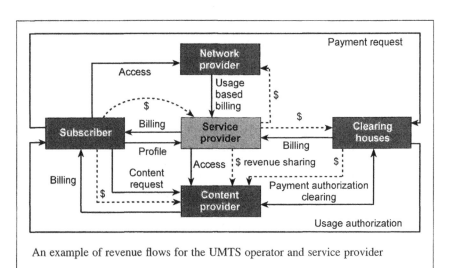

An example of revenue flows for the UMTS operator and service provider

15.2.2 What kind of revenue (and/or cost) sharing options

The range of revenue sharing options is broad. Let us assume only two players for a given service partnership, the operator and the content provider (although often an application developer, portal, IT integrator, and/or content aggregator may also be part of the equation). The operator may purchase the full rights to cellular delivery of the content. These may be exclusive (that no other mobile operator may offer the content) or non-exclusive. At the other extreme the operator may want explicitly to have no right and no affiliation with the content. An example of this could be adult entertainment.

Between those extremes, the operator and content provider could work on a one-time payment with limited rights, such as a limited amount of use, or limited amount of time that the content may be offered. The two parties could agree on revenue sharing based on billing of the service, such as for streaming music, or a usage-based revenue share, with a split based on time used with the content (e.g. games), the number of clicks or page views (e.g. serial content), etc. The venture may have particular risks on one side or the other, and part of the revenue sharing benefit or burden may be shifted to cover the risk.

For example, a new pop-culture-based cartoon character for example relating to a movie which has not yet been released, might have a certain minimum usage level defined before it would generate any revenue sharing to the content owner, to offset the development investment which the operator has to take just to enable the content. For various services there can be ceilings and a floor on what is paid to whom and under what conditions. If the concept is strong enough, the operator and content provider could set up a joint venture and split its profits, as is happening with m-banking, m-advertising and the music distribution industries.

15.2.3 What level of revenue sharing

When examining an industry soon to be worth a trillion dollars, certainly the guidelines on what to expect in revenue sharing are at least a 'billion dollar question' for any given industry. The different expectations of what any given partner might gain out of revenue sharing will vary greatly from one industry to another, and amongst players within an industry. A thorough examination of the existing and evolving value chains in any given industry must be examined as a starting point. To that the analysis needs to include the relative value propositions being brought to the new service idea, the relative strengths of the brands involved, and the extent of possible substitution both in terms of content (and/or application) and delivery channels.

Let us examine two examples. Selling books and selling maps. If we were to consider buying a new book by Stephen King, and considered purchasing it via the mobile phone, there is little benefit from most of the 5 M's. There is little benefit from the ability to purchase the book while walking around, and there are ample competitive delivery mechanisms, such as the bookstores, news-stands, Internet book resellers. There is little benefit from a sudden impulse purchase, as the book cannot be consumed in a few minutes anyway. There is not much added benefit from personalisation that has not been already achieved by for example the Internet bookstores like Amazon. There is little natural benefit to pay for a book by phone, as you typically would not consume the book on the phone anyway. And machine automation would not yield much if the other four of the 5 M's are not in effect, as the purchaser is a human anyway.

So if you want Stephen King, there will be no real substitute, and definitely in buying a book, the Vodafone, Orange, mmO2 and T-Mobile brands are immaterial in superseding the novelist's name (i.e. brand). Stephen King's publisher can easily insist on a similar distribution deal as exists with other book delivery channels. Here the network operator would have little bargaining power to insist on a better deal than what Amazon has currently.

Selling a map the situation becomes quite different. There is little brand identity with maps. We do know some tourist guides, such as *Fodor*'s and *Lonely Planet Guides*, etc., but for specifically the mapping companies, few common consumers feel strong emotional affiliation with one map above another. The relative power of the operator brand becomes more significant. If this is the map recommended by Vodafone or Orange, etc., it is likely to be taken as more trustworthy and accurate than just any old map in cyberspace.

In selling maps, the operator can bring several benefits from its network. The obvious first benefit for anyone needing a map is location. Get me the map for here, where I am, or often, the map which I will need where I am going. Here the operator can deliver information which the map developer cannot gain from sources other than the mobile operator. The digital TV company and the fixed Internet ISP will not know if Tomi Ahonen has flown to the UK or is still in Hong Kong. The customer database can yield considerable extra info for repeat customers. For example if last time in Switzerland I asked for the mapping software to show me Italian restaurants and cash machines which accept MasterCard, then the next map in Tokyo can automatically offer to show me those same requirements.

This example shows that if we approach two similar items sold in the same outlets today, books by Stephen King and tourist maps, the two might behave very differently as mobile services, and the operator and content owner could expect a different proportion of the revenue share, based on the added utility, availability of alternates, and strength of the relative brands.

Each industry will need to be examined and the role of mobile services needs to be explored. Several industry-specific reports exist which examine the potential role of the players in the emerging new mobile service market. One such analysis is in I-Date's Web Music report which includes the value chain for the music market.

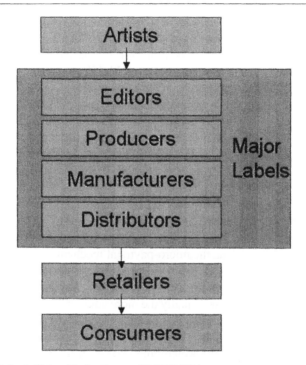

Music Market's Value Chain. Source: IDATE 2001.
As more music goes online and is downloaded electronically as "bits" of data the existing Music Value Chain will evolve and become more efficient and streamlined. The role of retailers and distributors will change and some artists will develop direct channels to their fans. This will only happen however when Digital Rights Management that protects Artists' copyright is introduced.

15.3 Main factors influencing split in revenue share

There are several main factors which have an influence on determining the ratio in revenue sharing. The only common denominator with mobile service partnerships is that the mobile operator tends to be one party. But the other party or parties can be any one or several of content providers, application developers, system integrators, portals, banks, advertisers, etc. Therefore, this discussion will look at the factors from the angle of the mobile operator.

Exclusivity is the first determinant. If the service needs a mobile operator for the service to function, there is a considerable amount of exclusivity as most other operators, such as cable TV, fixed telecoms and ISPs are not viable options. One example could be services delivered in real time to a moving car. Even further, there can be exclusivity among mobile operators, depending for example on a given technical standard, the availability of mobile phones and other terminals, geographic cellular coverage, etc. The fewer there are viable options, the more the mobile operator could expect to get as its share of revenue sharing. On a general level BWCS Mobile Matrix released a breakdown of revenues among players in the value chain.

The existing value chain or value chains are a significant determinant. In many industries the delivery channel has introduced digital delivery and these are likely to have the mobile Internet delivery option and its portion of the overall value chain relatively well defined. The mobile operator's share would be similar to what has already been introduced. If the industry itself is still in the early phases of conversion to digital delivery, then there is much more latitude for determining what is the mobile operator's share.

The extent of the value-add by both parties is of course also important. For the mobile operator, the main elements of direct value-adds are location information, access to user data, the micro-payment mechanism, and the ability to connect to direct communication systems, mostly voice and messaging. The more of the operator value-adds are used in building the service, the bigger the operator's share can become. Equally the more value-add elements the partner brings into the service, the more that partner can expect to get in its share.

On-screen location is another key factor. The ideal position for accessing a mobile service is right at the welcoming screen on the mobile phone. With very little total available space on the phone screen, and with thousands of potential services vying for that location, the operator is in a strong position if the service is set to be close to the top of the screen. The service can also be built to be behind one or two keystrokes, which would have a similar effect. The higher the service is placed near the top of the opening screen, the more the operator could expect to get in its share of revenues.

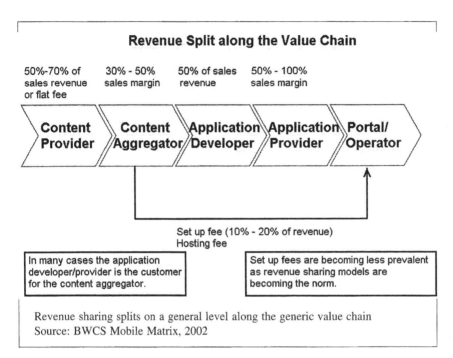

Revenue Split along the Value Chain

50%-70% of sales revenue or flat fee	30% - 50% sales margin	50% of sales revenue	50% - 100% sales margin

Content Provider Content Aggregator Application Developer Application Provider Portal/ Operator

Set up fee (10% - 20% of revenue)
Hosting fee

In many cases the application developer/provider is the customer for the content aggregator.

Set up fees are becoming less prevalent as revenue sharing models are becoming the norm.

Revenue sharing splits on a general level along the generic value chain
Source: BWCS Mobile Matrix, 2002

Finally the relative strengths of the brands are a factor. Brands will need to be considered always in the context of the given service and to its intended target audience. Tony Hawk might have little overall brand awareness when considering entertainment brands or even gaming brands, but among skateboard gaming Tony Hawk is one of the best-known brands. Depending on whether the intended service addresses skateboarders, the brand can be strong or weak.

15.3.1 Rules of thumb

Three simple rules of thumb can be taken from the digitally converging world. In the credit card industry the credit card companies tend to have commissions of between 2 and 4%. If the mobile operator offers no added benefit whatsoever from location information, subscriber data, urgency/time-liness, and only provides the billing convenience and takes a risk replacing a credit card company, it is reasonable to assume that the operator would take at least the same amount as the credit card company. As most credit card companies have minimum payments, if the operator wants to offer micro-payment options (payments with a value of less than 1 dollar), then the

15 Vignettes from a 3G Future: Back Seat Driver

My colleagues at work introduced me to this game. We live in a large city and the traffic congestion is often very bad. So if we have to take a taxi somewhere, we each start the real-time driving game and race the taxi driver – and each other – to the destination. The game is location.aware and uses the real traffic information. The real trick to winning in the game is to knowing good shortcuts and timings of traffic lights, etc.

As real-time networked games become ever more realistic, they will start to incorporate real elements from the current world. This would include real-time weather, and for example for a traffic driving game, all real traffic information from congestion to traffic lights. Combined with the location information and with a driver interface of speed, braking and steering control and a three-dimensional representation of the city in question, the backseat driver game can be created.

operator could be expected to keep more than the minimum that a credit card company does.

The other extreme is splitting in half. While this may seem quite extreme, in many emerging digital services which need the operator's know-how and systems, for example developing a multi-user real-time network game, it is fair to assume that the operator gets a considerable part of the revenue due to its considerable involvement. For most combined service development, a 50:50 split would seem the other extreme. In this case the operator would provide location information, customer data, billing services to the partnership, as well as its branding and location on its portal. Also the operator would take an active role in the system development or integration involved.

A third rule of thumb is the I-Mode example. DoCoMo's I-Mode service is widely quoted as taking a 9% revenue share cut from the value of the content or subscription to the official I-Mode sites. If no better initial benchmark exists, this can be a good starting point. A fair approach could be that starting from 9%, the two parties could examine what is the relative merits that either partner brings to the partnership, should the revenue share be more or less than with I-Mode. From that the relative extremes could be about 3% as the absolute minimum and 50% as the absolute maximum to be the part going to the operator.

Network operators will need to set pricing for their various services and levels based on their needs. The content providers, application developers and any other parties wanting to get into revenue sharing need to look at their various options and consider what they would find acceptable with the operator. The content providers must be active in seeking the opinions from all players in their market, and use the bargaining power of playing one offer from one operator against the other to secure the best deal. Similarly the network operators must remember that in their portfolio of hundreds or thousands of services, there are rarely unique service offerings, and they too should be in contact with several candidates to find the preferred partners.

15.4 Keys to success

The chapter on the 5 M's (Chapter 5) showed how any content can be enriched when brought to the mobile Internet. But even the best service or product may fail in the marketplace. More typical of both the IT and telecoms industries is that success comes suddenly from unanticipated sources. A good example is DoCoMo's I-Mode service, which at launch expected to have its success coming from businessmen gambling on sports. Most of the revenues come from young people playing games or downloading cartoon characters and other such youth-oriented entertainment. This has obviously been a surprise in the Japanese market. Similarly in Europe the counterintuitive SMS (Short Message Service) text messaging service was early on not seen even as a viable business by most European operators, until some of the early data started to filter in from early adopter countries such as Finland. Now practically every mobile operator offers SMS text messaging.

New success can mimic success in the entertainment, gaming and pop culture areas. Content built around TV shows with large audiences, such as *Who Wants to Be a Millionaire*, *Survivor*, and *Weakest Link*, seem likely candidates to succeed in the areas of mobile Internet content. The same would go for *Harry Potter* and *Lord of the Rings* movie tie-ins, as well as anything built around the latest pop music stars. The content can take on a multitude of formats, from direct content (songs, video clips, foreign language versions, books, etc.) to news, gossip, backgrounders and analysis, and on to games, cartoons, merchandising, etc.

Remember also that successful content can appear seemingly out of nowhere and grow fast to remarkable popularity. Many a popular phenomenon can also disappear fast, as the tamagochi phenomenon can prove. Telecoms operators are accustomed to building services which have a lifespan of decades; many popular content might have a lifespan measured in months, even less.

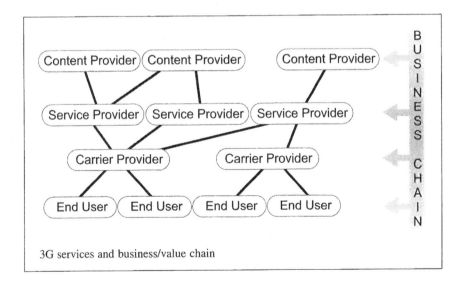

3G services and business/value chain

15.4.1 Profitability

Often considered a black magic science, shrouded in mystery and secrecy in telecoms, the most important single attribute of a new service is of course its profitability. As mentioned before, telecoms operators are not known for understanding the profitability of their products well. Here is one area where the partners need to assist the partnership by keeping the profitability of the partnership venture in focus.

Profitability needs to be calculated for every separate service as well as any given service bundle. When considering a large new portfolio of new mobile services, a mobile operator has to give its controllers and product managers clear guidelines on how to determine profitability, so that the profitability calculations and measurements for the whole product range are compatible and comparable. Often this includes a management indication of network costs including both operating costs and capital expense. With most telecoms operators this type of information is held to be extremely sensitive and available to a tiny part of senior management. In handling the profitability of a portfolio of mobile services it simply is not acceptable to hide the real costs from those whose job it is to manage the services and oversee their sales.

15.5 Let's stick together

More new mobile services will be made with partners than alone by any of the major players, including the mobile operators. For a partnership to work, both (all) parties have to find real benefits on a long-term sustainable basis. Revenue sharing will be a key ingredient to such partnerships and mobile operators will be on a steep learning curve to achieve the ability to manage such partnerships.

Some partnerships will end up not working and others will run a natural course and end. Most partnerships will have their challenges and will require work and effort. When the first problems arise the easiest path is to abandon the partnership and seek a new partner. While it may be difficult and at times frustrating to work with new partners, it is best not to abandon new relationships early, but to give them the extra effort in the early stages of new technologies. Crossing technical chasms are not unlike crossing rivers, and we are reminded of what Abraham Lincoln said, ''It is not best to swap horses when crossing streams.''

16

'Benjamin Franklin may have invented electricity, but it was the one who invented the meter who made the money.'

Earl Wilson

Business Case of 3G for the Operator:

Revenues, costs and profitability

Where is the money. Strongly simplified, all business cases look at revenues and costs, and see if after you deduct all the costs from all the revenues, there is any profit left. This chapter will provide an overview into the major parts of the 3G business case from a network operator's point of view. It is written with the author's exposure to dozens of real 3G operator business cases, and through discussions with 3G business case experts from leading global consultancies, leading investment banks, and industry analysts who specialise in telecommunications and 3G. It is a simplification and an amalgamation of various inputs.

The 3G operator business case as presented here is intended for those who are not in the network operator business to understand on a general level what is involved. For a deeper understanding of the 3G business case, please turn to one of the specialist publications on that subject.

16.1 Business case basics

In the 3G network operator business case on the revenue side operators get most of their revenues from callers, usually called subscribers, and most of this revenue is directly resulting from traffic, such as minutes of airtime or messages sent via SMS (Short Message Service) text messaging, etc. Mostly network operators also charge monthly fees for the subscription. Some revenues also come from other operators in the form of interconnect revenues. A significant new revenue source in 3G will be advertising revenue.

On the costs side, for cellular telecoms operators, it is typical to divide the costs into fixed network equipment costs (CAPEX – CAPital EXpenses) and operating costs (OPEX – OPerating EXpenses). In 3G in many countries the licence fee has become a very significant new component but in other countries a 'beauty contest' was held with at best trivial licence fees. And as new content-oriented services will emerge, an ever increasing need will be the various data servers which support the content that is consumed, for example the file servers holding today's newspaper in its small-screen mobile service format.

16.1.1 Service revenues

Revenues to operators come mostly from services. By starting from the existing services in 2G networks, the current biggest sources of revenue are from voice services (voice calls) and messaging services (SMS text messaging).

16.2 Market drivers for 3G

Several concurrent market trends exist today which create a strong business opportunity for the 3G network operator. The emergence of e-business and its related digitalisation of content is a major driver to bringing content into digital format(s). The recent failure of the Internet economy to sustain itself solely on banner advertising revenues is another driver, which has the content community looking for alternate revenue sources.

16.2.1 Terminal evolution

The terminals (mobile phone handsets) are growing ever smaller and lighter, with longer battery lives and better screens. The functionality of standard handsets of 2001 has many features which existed on top-of-the-line phones only 2 years ago. The next developments, already seen in some models, include the addition of music players on to the phone, combining the functionality of the Walkman and mobile phone; the addition of FM radio; the use of high resolution colour screens to enable viewing of pictures and images; and the incorporation of the digital camera to the mobile phone, producing a hybrid between the mobile phone and the snapshot digital camera. As the memory and storage ability of the handset keeps growing with Moore's Law, the near future mobile phone handsets will also be able to do many things we today will have a hard time imagining. (Gordon Moore, the founder of Intel, noticed that computing power will double every 18 months with the cost level of computing remaining constant.)

The 3G phone in a few years will easily incorporate many features of gadgets we consider carrying with us today. The 3G phone can be a multifunction device which has the ability to capture images and pictures, store and send them; entertain us with games and music; allow us to send and receive messages and e-mail; access the web; perform little daily routines like handling our calendar or do calculations and conversions; and of course still function as a phone. The early multi-function phones are likely to be somewhat bigger, bulkier and heavier than today's phones, but less so than combining for example the digital camera and the mobile phone of today. And as miniaturisation continues, together with Moore's Law, we will soon see smaller 3G phones than today's mobile phones, packed with a lot of functionality that would seem impossible today.

16.2.2 Standards harmonisation

The current incompatible mobile phone standards for digital cellular telephony, which can broadly be divided into three widely adopted classes – GSM (Global System for Mobile Communications), TDMA (Time Division Multiple Access) and CDMA (Code Division Multiple Access) – as well as several national variations and less common standards, are on evolution paths bringing about harmonisation and primarily only two remaining 3G standards – WCDMA (Wideband CDMA) and CDMA2000. WCDMA seems to be the clear dominant standard selected with most of the world's largest countries and cellular operators, but CDMA2000 has a strong group of carriers also committed to it. The result is that network infrastructure manufacturers and terminal manufacturers can focus more and limit their involvement in costly R&D across many standards. Also the fact that essentially only two 3G standards remain, ensures that most equipment makers in cellular network infrastructure or in mobile phone handsets, will be competing in this space, on the same standards. The wider range of suppliers is likely to promote technological innovation and competition of course helps bring prices down.

16.2.3 Subscriber growth

The single most dramatic trend leading into 3G, is the incredible year-on-year subscriber growth around the world in mobile operator networks. Practically all early predictions of the levels to which mobile phone subscriptions might rise, have been exceeded. Numerous countries have followed Finland's lead with mobile penetration exceeding fixed line penetration, and Finland with penetration around 80% in late 2001, is still showing no signs of cellular subscription growth slowing down. Now it is clear that many adults are starting to have two mobile phones, and predictions in Finland suggest mobile phone 'human' penetration will reach approximately 120%–130% of the population. Separately there is the matter of machine penetration.

The possibility to have machines connected to the cellular network and transmitting traffic is also a scene from near history science fiction. But early concepts such as the intelligent refrigerator calling your supermarket to order more milk, have enabled a lot of creative thinking around how gadgets and devices can communicate, and many of them via the cellular network. Some of the most obvious ones are related to automobiles and metering devices generating 'telematics' traffic which was discussed in the chapter on Machine services (Chapter 10). For every one of these machines, there will be a SIM (Subscriber Identity Module) card which will seem to the network just like a human on a mobile phone, which also has a SIM card.

SIM card penetration is expected to reach a multiple of human penetration, typically telematics experts estimate machine penetration to have 2 × or 3 × the subscription amount of humans. Of course machines will not behave like humans, they do not spontaneously call their best friend to talk, nor will they feel the urge to call mother on Mother's Day, nor to go surfing on the *Playboy* website. Machines tend to put traffic in very clearly predictable patterns and only to pre-determined locations and addresses.

Regardless, the recent phenomenal growth in mobile phone subscriber numbers, and the signs that the growth will still continue into the near future, combined with new 'subscriber' growth in the form of machine SIM cards, is a trend that cellular operators are eager to cater to.

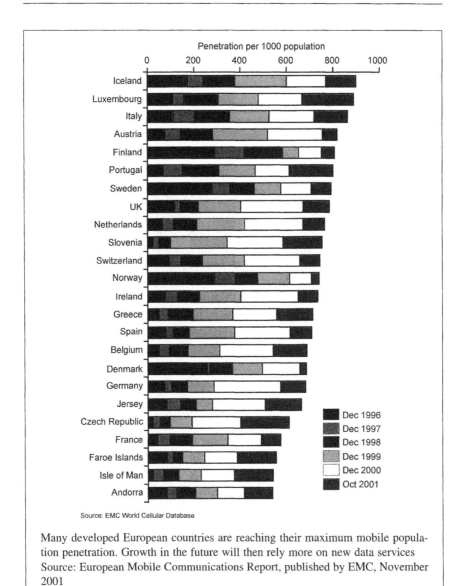

Penetration per 1000 population

Source: EMC World Cellular Database

Many developed European countries are reaching their maximum mobile population penetration. Growth in the future will then rely more on new data services
Source: European Mobile Communications Report, published by EMC, November
2001

16.2.4 Traffic congestion

The unexpectedly strong total subscriber base of second generation cellular
subscribers, and the eagerness of those subscribers to put phone traffic on to

the network, has produced a massive overdose of telecoms traffic on 2G networks. This has greatly exceeded the early specifications of the maximum load that these networks were designed to carry. The result has been that in most major cities especially in Western Europe with high mobile phone penetrations and usage, the traffic congestion has become a very significant problem, especially at peak hours.

In London, with a long history of GSM, and with four competing networks, and the network operators with very similar market share, it is still quite common to drop out of the network and have congestion problems. So the problem is not that the technology is new, nor is it a lack of competition and 'bad service by the incumbent' – but a very real issue of the current 2G standard being stretched past its ability to deliver. The traffic congestion is actually getting progressively worse, not only with ever more subscribers connecting and with existing users more and more willing to talk longer on the mobile phone, but now with the introduction of higher bandwidth-hungry services and data connection services on WAP (Wireless Application Protocol) and GPRS (General Packet Radio System).

There is a clear appetite by cellular network subscribers to put traffic on the networks. And the current technology cannot even serve the current population adequately. With new services being possible in the near future, there is a clear need for higher bandwidth-capable, capacity-optimised networks with high spectral efficiency. That is what 3G is. 3G will not by itself bring many radically new services that cannot be done on 2G or 2.5G technologies, but 3G will allow very high cost efficiency in delivering those services.

16.2.5 Digital convergence

Worldwide there is of course the general trend of digital convergence, by which all communication and content are proceeding to converge. Voice and data, mobile and fixed, content and delivery area all converging. The book, newspaper and magazine printing industry is going digital. The television and radio broadcasting industry is going digital. The fixed telecommunications and mobile telecommunications industries are far along converting from analogue to digital. The gaming industry is getting ever more digital, connected and mobile. The motion picture industry is going digital. The list goes on and on and on. This convergence does bring about disruption and opportunity. A natural 'end state' will be the 3G environment which will be mobile and fixed, data and voice, and fully digital. It will be a major area where the convergence will be experienced.

While all these industry changes are taking place, mobile operators are in a key position to exploit new business opportunities brought about by the mobile Internet. The *key assets* of the mobile operator in this 3G business model are the micro-payment billing infrastructure, a large end-user base, an established mobile brand, the users' location information, established dealer channels and, naturally, the mobile network infrastructure itself.

16.3 Service revenues

The services that the 3G network operator plans to run can be divided into many different groupings and analysed many ways. Some of the more commonly used divisions are by type of service such as voice and non-voice (often called voice and data); by technology used to deliver the service, i.e. 2G and 3G services; or by customer types, e.g. business and residential services, or by the entity paying the bill, i.e. subscribers vs advertisers, etc. For the purposes of this overview, a simple division by type of service will be used.

16.3.1 Voice and its evolution

The basic service on all current cellular networks is voice calls. Voice accounts for anywhere from 85 to 97% of the network's traffic and 80 to 95% of the network's revenues in 2001, on all the major digital standards TDMA, CDMA and GSM. Operator business cases expect that voice will continue to play a big part in the future of cellular telephony. Voice services will continue to grow in traffic for the next 5 years and beyond. As voice traffic grows, with the emergence of more competitors into the 3G network operator competitive field, the general assumption is that voice services tariff erosion will continue, and that it will escalate faster than voice minutes growth, and thus tariff erosion will exceed the rate of voice traffic growth, producing a gradual decline in voice revenues per user.

A new form of voice traffic and revenue will emerge as what is called 'rich call' services. Rich calls are described in detail in the Me chapter of this book (Chapter 8). Rich call services will provide higher margin revenues to 3G operators and are expected to bring an ever growing proportion of total voice revenues to the operator.

16.3.2 Messaging – SMS and multimedia messaging

The recent major source of revenue especially in the GSM operator environ-

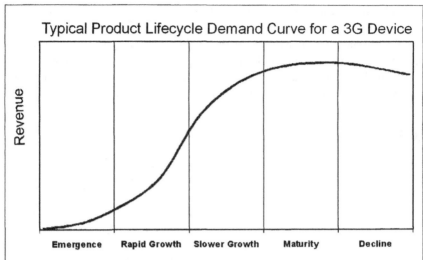

Typical Product Lifecycle Demand Curve for a 3G Device

Revenue

Emergence   Rapid Growth   Slower Growth   Maturity   Decline

Typical product lifecycle demand curve for a 3G device according to the UMTS Forum report Number 9

ment is the emergence of SMS text messaging. SMS and variants and extra value services built upon SMS count for anywhere from 5 to 20% of revenues with GSM operators. SMS has been growing dramatically and its use is not slowing down, not even in the first adopting countries such as Finland. SMS text messaging alone is expected to witness price pressure from independent messaging engines and sources, such as fixed Internet service providers like Yahoo and AOL, who are likely to offer such services also as mobile portals. And the price levels of SMS messaging are likely to see some price erosion over time.

Still for the next few years, SMS traffic is usually seen as continuing to grow, at least for the next 5 years or so. And SMS price erosion, even if strong, is not likely to cut strongly into the 3G operator's business case within the next few years.

A new form of messaging services is multimedia messaging. Much like current e-mail consists of ever more attachments going with the text files, so too very soon multimedia attachments will be part of SMS text messaging. Sending pictures, video and audio clips, game scores, etc., with messages will generate higher value messaging traffic. Multimedia messaging is typically expected to deliver about as much revenue in 5 years as plain SMS text messaging.

16.3.3 Data access services

Another area of the 3G business case which is already familiar to the current network operators, and shows up as a significant part of their service revenues, is data access. Currently this mostly means connecting laptops to corporate LANs (Local Area Networks) and WANs (Wide Area Networks) and a myriad of intranets, extranets and the Internet.

Data access services in 3G will continue to include the current services with their usage expected to grow dramatically. Data access services will also include the data traffic related to 'value-add' or 'portal' services related to m-commerce. This analysis makes a distinction between the data transmission part of the value-add traffic, which typically the 3G network operator is unlikely to share with content providers, and the part of the traffic which is strictly value-add, and which mostly the operator would need to share with the various content and application providers.

16.3.4 Value-add services (also often called portal services)

Value-add services are mostly a new type of traffic and content, delivered via the mobile devices. The Japanese market has exploded with new ways to consume digital content on the three major telecoms networks, NTT DoCoMo's I-Mode being the best known as the market leader, but also KDDI and J-Phone having impressive subscriber growth and service usage. Similar trends are emerging on European services. Portal services include information push and pull services, i.e. newspaper and yellow pages as examples. They include m-commerce and its various incarnations from buying books online to mobile banking, etc. The value-add services also include games and entertainment services, such as trivia quizzes and daily cartoons.

Typical of all value-add mobile data services is that the content provider is usually a well-known outside content provider, and the network operator will be compensated a 'standard' amount for the transmission of the data based solely on how much the actual load is on the network. A second part of the payment to the network operator is a 'commission' which most often varies, based on the value of the content. The commission can be as little as 4–5% of the value of the content if the commission replaces current credit card transactions but nothing else, and essentially only provides charging and billing. The commission can be as much as 50% if the 3G network provider brings in a lot of value-add benefits that the content provider wants but cannot easily get elsewhere, such as location information, personal information on the

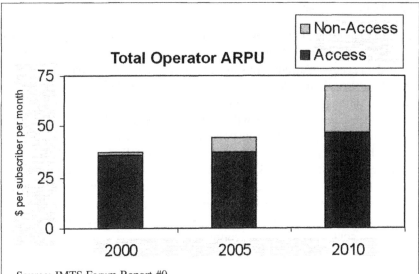

Source: IMTS Forum Report #9.
Average Revenue Per User (ARPU) has decreased in the past 3-5 years driven mainly by the increase in low usage prepaid subscriptions but ARPU now appears to be levelling off in many markets. Operators with the right strategy can look forward to increasing ARPU levels.

subscriber, the portal bundling benefits, and brand awareness of the network operator, for example as a preferred partner, etc.

The content commission revenues to the 3G network operator for the value-add or portal services are expected to bring in between 10 and 20% of the total revenues in 3G services. Note that these same revenues bring in about the same amount in data transmission revenues mentioned above.

16.3.5 Machine-to-machine or telematics services

Telematics and other machine-to-machine services will bring in revenues, but typically a small amount of the total. While there is likely to be a large population of 3G devices which will only communicate with other 3G devices, such as metering devices and automobile systems, these usually transmit small amounts of data only, and they can also be configured to transmit in background transmission classes and during off-peak times. The purchasers of machine-to-machine traffic tend to be large corporations for a fleet or vast array of these devices, so they will be using their bargaining power in securing a low cost per bit transmitted. Machine-to-machine traffic typically accounts for a few per cent of the total traffic revenue for a 3G operator. Automobile telematics typically are considered to be particularly suited for incumbents who have already built networks to cover the highways, motorways, freeways, autobahns and autostradas around the world. The newcomer or 'Greenfield' 3G network operator is not expected to be strong initially in providing telematics services.

16.3.6 Advertising revenues

Advertising revenues have a wide dispersion in various industry reports making estimates of the future. Operator revenues have been estimated anywhere from less than 2% to over 20%. Of course the overall acceptance and attitudes towards advertising is a factor, explaining partly why so much of the advertising innovation in 2G and 2.5G mobile services is coming from America and Japan. There is a lot of scepticism towards mobile advertising in countries where it has not yet been seen in major campaigns, but that is similar to the initial doubts about TV advertising and advertising on the radio in the 1980s and 1990s in many parts of the world.

The important part in mobile advertising for the 3G network provider to understand, is that advertising is an art form, and its best artists are particularly demanding employees. Even if a 3G network operator was able to initially create some exciting mobile advertising with its in-house marketing staff, very soon the creative input into successful 3G advertising and promotion will come from the true masters of it, who all work for the major advertising companies. Even more in advertising, 3G is not at all about technology, it is all about services and content. The best at making advertising content is the advertising industry. The long-term success for 3G network operator advertising revenues is becoming friends with the advertising industry and working with them. It does mean that the 3G network operator will have to surrender a lot of the revenues to the creative talent: the advertising agency.

16.3.7 Interconnect revenues

Interconnect revenues are part of the more secretive elements of the network operator community. A few major points should be discussed. First of all, interconnect is always a two-way street. There is interconnect *revenue* when I terminate a call which your caller placed to someone in my network. But there is interconnect *cost* when my subscriber calls someone in your network and you complete the call. Interconnect is paid to the party who delivers the connection, and it is not uncommon that more than one party receives interconnect revenues for a given call or connection. For mobile operators, interconnect tends to be a net-positive account. In other words the mobile operators tend to receive more interconnect revenues than they pay out in interconnect costs. This is because of the costs of delivering a call in a fixed network is usually a lot cheaper than the cost of delivering a call in a mobile network.

As mobile substitution continues, and more and more people start to abandon their fixed line altogether, as has been seen as a trend in Scandinavia already for several years, then the number of users in fixed networks will diminish, and the proportion of users on mobile networks will continue to grow, approaching 100%, but unlikely to reach that in the next few decades. As the proportion of calls originating from fixed networks will decline, so too will the net positive interconnect. And the mobile operators will eventually notice that almost all of the interconnect traffic is between mobile operators, whose interconnect termination charges will cancel each other out.

Over the next few years, interconnect will continue to be a significant part of 3G network operator revenues. Typically operators have interconnect at less than 10% of their revenues today, and it will still be at least 5% in 5 years.

16.3.8 Does my phone bill double?

In the early days of mobile calls, and in countries where receiving party pays, such as the US today, the concept of the ARPU (Average Revenue Per User) is directly related to the end-user phone bill. For American readers: in Europe the operators have already moved beyond this concept as they have calling party pays, and thus many have their mobile phone for primarily receiving calls, net-receivers of calls, while others are net-callers.

With the advent of 3G the ARPU will not be the same as the end-user phone bill. If one looked at the analysis above, and saw that the ARPU is doubling, it would be easy to conclude that your phone bill will double. And then to think of your own growing mobile phone bill, and reach the conclusion that very soon into the 3G business case the network operators will simply hit the wall that the end-users cannot pay for more.

It is a fair conclusion, but based on a faulty premise. In 3G while the ARPU may double, the end-user phone bill for residential users will not. In fact, the end-user phone bill for current services – mobile phone calls and SMS services – will actually diminish. How can it be? Several factors contribute to this. First is that several of the new services, especially the high-cost ones will be sold to businesses and business users. For example the secure access to the corporate intranet is important for corporations with intranets. Few homes have a need for a secure private intranet. Video calls in the form of video conferencing will be among the most costly of new services, and they will again be marketed to multinational businesses, where a video conference is much cheaper than executives flying to meetings around the world. So several of the new services will be targeted at business users, and the residential phone bill will not see an impact.

The operator will get additional revenues from many completely new sources, which are not billed to the mobile phone user. For example telematics traffic relating to commercial metering and instruments would produce traffic into the network, and revenues to the operator, but this new money would not come from the pockets of the mobile phone user, at least not directly. Similarly the operator would get significant new revenue sources from m-commerce transactions and advertising and sponsorship revenues. That is why even though operator ARPU doubles, the phone bill for the consumer does not.

One last thought needs to be kept in mind about operator revenues and ARPU. ARPU is not the same as operator revenues. ARPU has numerous deficiencies, in particular as it was a measure intended for an industry with only one service (voice) and now faces service portfolios with thousands of services. ARPU will distort revenue analysis unless it is combined with subscriber numbers. A vital sanity check is overall revenues. A mobile operator can be extremely healthy and have a declining ARPU figure. Equally an operator can have a growing ARPU and be in trouble. Always look at subscriber numbers and total revenues in addition to ARPU.

16.4 Capital expenses

The costs of the 3G network can be divided again in many ways. This analysis will use the division to network set-up costs, capital expenses (CAPEX), and network running costs, operating expenses (OPEX). For the purposes of this analysis the costs of acquiring the licence are allocated to the CAPEX side, even if in some countries the licence fees are paid over several years, etc.

16.4.1 Network dimensions differ radically

It should be noted that the capital expenses of a 3G network operator can easily vary greatly from one operator to another. A 3G network is a cellular radio network. To build it the operator (carrier) needs to deploy a grid of base stations in a pattern into what is called cells. Cells are typically kilometres in size. So a typical cellular network might contain several thousand cells. Immediately the impact of geography will determine greatly what is the cost of setting up a national network. In a densely populated and relatively small country, such as Hong Kong, Singapore or The Netherlands, or the state of Massachusetts in the US, it would not take very many cells (and thus base stations and sites) to set up a cellular network. And if the country is sparsely populated such as Sweden, Canada or Australia, or the state of Montana in the US, then it would take many times more cells (and base stations and sites) to build up a network.

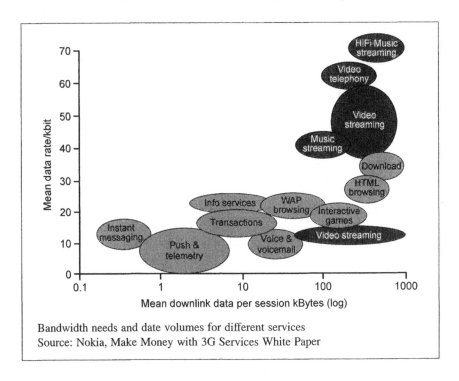

Bandwidth needs and date volumes for different services
Source: Nokia, Make Money with 3G Services White Paper

Secondly there is the concept of geographical coverage and population coverage. Typically cellular networks are built first in cities (dense urban areas) and then in the outskirts of the cities, the suburbs (urban areas) and into ever less populated areas (rural areas) but usually up to a point of practical limits leaving out uninhabited or very sparsely populated areas such as major forests, mountains, etc. Thus the actual coverage built will be similar to the population patterns within a country, but it means that there are drastic differences in covering countries of similar size, or if the operator intends to have a coverage footprint of a different size to its competitor. To illustrate by real examples. To cover about 80% of the population of the US would take several tens of thousands of base stations. Of course the population of the US is spread relatively evenly across the country. To compare countries of a similar size, for a country like Australia with almost all of its central area being a barren desert, or Canada where most of the population lives within 100 miles of its southern border to the US, to achieve coverage to 80% of the population would only take about a thousand base stations.

16.4.2 Network sharing

Another aspect is the concept of network sharing. Network sharing can have a dramatic impact on the initial rollout costs of the network. There are costs to the management and differentiation that can be achieved on a shared network, and upon traffic congestion, a shared network delivers no value to the sharing partners. Hence many network sharing schemes include plans to dismantle the sharing if traffic reaches predetermined levels. In Sweden, Chalmers University conducted a study which indicated the levels of savings possible under various scenarios.

There are cases where network sharing is not viable and cases where it makes very strong business sense. For example in very sparsely populated areas, if two operators share networks, they can achieve dramatic savings in radio network dimensions when compared with a single operator building the same coverage alone.

Suffice it to say that the summary given here can provide indications of the costs involved, but in individual cases the costs can easily be over double these, or less than half those discussed below. But one should also remember that as will be illustrated later, the operating expenses are proportionately much greater than the initial capital expenses, and in the operating expenses one can make many generalizations across 3G network operators in most countries.

16.4.3 Licence cost

The licence cost varies greatly. In some countries like Finland and Sweden, the governments all but gave the licences away and they are no factor in

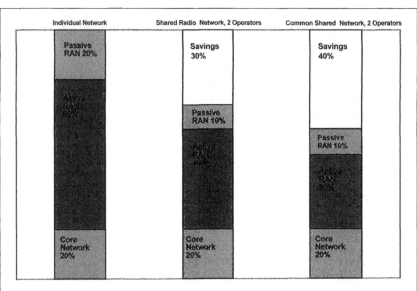

Savings from Network Sharing, Chalmers University of Sweden Study 2001

the business case for the 3G network operator. In other countries like the UK and Germany, the licences went up for auction, and the auction process attracted very heated interest and competition, pushing the cost of licences remarkably high. Some analysts have calculated that in those cases the licence cost is of the same magnitude as the cost of building the network.

Of the individual cost elements of the 3G operator business case, the licence cost has the greatest fluctuation, and definitely can be the single largest cost element in contested auction countries. The licence cost is also a matter of public record so it can easily be determined for any given 3G network operator in any given country. As the amount fluctuates so greatly from being the biggest single cost element to being totally meaningless, lost in a rounding-off error, it is not practical to include any one number in this sample analysis. Its size and impact needs to be noted in any analysis of any actual operator's 3G business case.

16.4.4 Radio network

The radio network is the largest component of a 3G network, both in size of equipment and in cost. The radio network includes the antennae, the base stations, the sites, etc. To put it in perspective, in a typical case between 60 and 70% of the total network cost is directly related to the radio network.

The single biggest cost in the set-up of a radio network is actually not radio networking equipment, but rather the cost of acquiring the sites to house the equipment. A typical base station is the size of a large refrigerator or two, and it needs a room relatively near to the location of the antenna. The antenna is placed on a mast and the masts' heights are governed by local regulations. The antenna height is one factor in cell sizes, so when antenna height is low, more cells are needed and thus more sites and equipment. But high antenna masts are considered an eyesore. There is also the need of more power to transmit over larger distances, especially in penetrating into buildings, so antenna height alone cannot be used to create large cells. The power requirements and the topography of the surrounding landscape will also be a big factor in the size of the cell that can be achieved. In many cases national legislation is limiting the power output of cellular transmissions, setting limits to the dimensioning.

16.4.5 Site acquisition cost

Still, in a typical city with cell sizes of a few kilometres, sites will be needed in every part of the town, and thus many locations in every city. The cost of acquiring suitable sites, where the location is good for an antenna and fits the cellular transmission grid, means that individual site costs in downtown areas can be considerable. Imagine walking into a fully staffed bank headquarters building downtown and trying to negotiate the size of an executive's office for use of your network base station equipment, and get access to that location as well to do maintenance, etc. The site acquisition costs are the single largest cost element for setting up the 3G radio network. Note that often the sites may need electrical work to provide power and air conditioning, etc., they might need floor strengthening work, special telecoms cabling work to connect the site to the rest of the network, etc.

16.4.6 Base stations

A typical 3G network built on the most dominant standard, WCDMA, will reach a given geographical footprint in a very similar network grid as GSM

1800. So for a typical large European country, the 3G network operator would be looking at about 10 000 base stations, in a grossly generalised simplification. For CDMA2000, with its significantly different performance ratios, costs, and network dimensioning rules, the actual size of the radio network is similar. For the overall 3G business case, the revenue side is that much more significant, that the monetary impact of the technological choice in the whole radio network can be lost in a rounding-off error. This is not to belittle the costs of the radio network, and millions of dollars. It is only to remind the reader that if the revenues are wrong, it makes no difference how perfect the network is.

The 10 000 base stations for a Britain, Germany or France is a good round number to keep in mind, in that we are not looking at 200 base stations, nor are we looking at half a million. When you imagine about 10 000 'refrigerator sized' electronic devices brought to their sites and set up and connected to the network, you can imagine that a typical network build-up is measured in years, not months. The build-up will typically start from the largest concentrations of people, the cities.

16.4.7 Core network

The core network includes the central control functions such as service creation, charging and billing, provisioning, etc., and the backbone to deliver the bulk traffic from parts of the network to other parts, as well as to provide interconnect points to other networks. The core network is seeing a major stage in evolution going from circuit switched networks, to packet switched networks, with the introduction of GPRS elements. The whole core network typically costs between 20 and 30% of the total network cost. The biggest elements in the core network are the service platform, billing systems and the transmission network for the core traffic. In some developing countries the cost of the high capacity telecoms trunk network can rival the cost of all other parts of the core network combined.

The fastest growing element in the capital expenses overall, relates to the new services. It is the platform costs of various service creation and management platforms, which will evolve probably close to the speed of the IT industry's rate of growth, roughly needing to be replaced every 3–4 years, and the replaced equipment inevitably having so many advanced features that it costs more than the equipment it is replacing.

16.5 Operating expenses (OPEX)

The operating expenses for 3G networks will be similar in their types to operating expenses of mobile networks today. The single biggest operating cost element is marketing cost. Easily a fourth of all OPEX is marketing cost. The marketing cost includes the handset subsidy where relevant, which in some countries is the biggest individual marketing cost. Of course in other countries handset subsidies are totally forbidden, and in many countries the operators have noticed that subsidies lead to skewed behaviour by customers, and operators are trying to move beyond using subsidies as a marketing tool.

The biggest growing operating expense relates to the new services and their management. The 3G network operators will need to invest in staff to manage the numerous partnership arrangements relating to the 3G content on their network. It means more attorneys to draft and review and update contracts, more product managers to manage groups of products and watch over their profitability in an environment of shared revenues, more partnership managers to act towards the key content and system partners to smooth out any concerns in those volatile directions. The costs of managing the technical network, maintenance staff and network management staff, is not likely to grow significantly compared with current 2G networks.

16.6 Profits or losses

There have been many profit and loss calculations in the public domain on the 3G business case, from the very rosy 4 year paybacks, to the 'never will pay back' and 'minimum 12 years' and so forth. There are many variables which are unknown. At one level one could say that costs and profits are the inverse of network quality, as illustrated by Kaaranen in the book *UMTS Networks*.

At the 3G operator business case level there is also a wide variance. I have personally been involved in the business case calculations for several operators, and have discussed the business case in depth with the operators themselves, Nokia's business consultants that I have supervised, and with several outside experts on 3G, such as 3G consultancies, investment bankers and industry analysts. From those discussions, I can summarise that the common feeling is that for an incumbent operator the 3G business will break even in about 4–7 years depending on licence costs and the projected market share.

The payback period for newcomer networks, so-called Greenfield operators is between 7 and 10 years.

How reliable is that general consensus? The cost side of the equation can be modelled with considerable accuracy, but unfortunately for the accuracy of

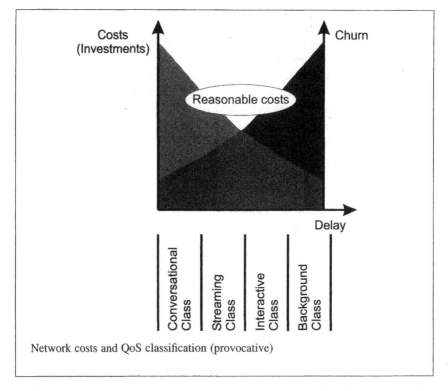

Network costs and QoS classification (provocative)

the forecast, the cost side is not the dominant one as was described above. The revenues dominate the business case. And since more than half of the revenues are expected to come from services which do not exist today, this business case is based on a lot of belief. The operators who do not really know whether advertising will be 3 or 11% in 2005, are very willing to quote a general number which has been found to be accepted. That is why so many of the business cases seem to be harmonising on to the revenue projections stated in Report #9 by the UMTS Forum.

In a period of technological change, at the advent of new business models, and with no historical data to use, there is considerable uncertainty with public statements. One has to keep in mind, that when nobody knows for sure, if someone 'reliable' quotes a given number, then it is easiest to agree with that, rather than quote another number which is significantly different. We will not know for sure until we see how each of the new services takes off – or not – in the marketplace. Keeping in mind the hockey sticks, we can then start to enter real numbers into the models, and get a more accurate picture around the winter of 2003–2004. Up to then, the best of the models provides only an educated guess about the future, nothing more.

16.7 Sensitivity analysis

The biggest single factor in the business case is the development of the revenue streams of the services, and thus ARPU. A small change in the ARPU will result in a dramatic change in the overall payback periods, net present values, internal rates of return, etc. It is vital for the 3G operator to understand the services and how to nurture the traffic growths, and understand the value chains, and ensure the profitability of any given service. It is critical that the 3G operator is early in any target service area, and that the operator is responsive to changes in the market.

If fast revenue creation is the most important factor an operator has to consider, the then second is clearly operational efficiency and cost control. Two-thirds of 3G capital expenditure will be in the radio access networks, so this is where the largest early savings can be achieved. Then the focus should shift to OPEX and especially in optimising marketing cost-effectiveness and minimising the drain from new IT service costs involved in managing new data services.

Naturally, each business case is different and this is why operators have to use sophisticated business and techno-economic modelling to run scenarios and plan their strategies for network rollout and service creation. The key to success in operating a 3G wireless network is to understand what content, what applications to develop and with what business models.

16.8 Bring the noise

This chapter has examined the business case for 3G. The case depends on very many variables and with the complexities involved a whole book could be written just on the business case for any given operator. In fact such 'books' *should* be written at those operators in the forms of strategic business plans, network designs, marketing plans, etc. But for the general public and as an overview to anybody involved in 3G, this chapter has attempted to show the importance of services and revenues, what is the ARPU and how it can be used, as well as how it is likely to evolve as a measure.

This chapter has also looked at the capital expenses and operating expenses involved in the 3G business. Before one can be successful in 3G, one has to understand its nature. The reader should also seek further opinions of opposing viewpoints before forming a final conclusion. But while the investments are huge, the payout is expected to be even better; that is the nature of investing in a new technology. As Peter Drucker said ''Whenever you see a successful business, someone once made a courageous decision.''

17

Money Migration:

Know the streams

The digital content and communication market will be growing and most parts of it will be growing much faster than the overall economy. There are several predictions of growth rates for the different parts of the whole which can provide guidance on where opportunities lie. The general growth trends hide very significant shifts between the different parts of the whole. This chapter examines how usage, revenues and profits are likely to migrate from one part of the overall telecommunications market to another.

17.1 Four major parts

The Money Migration theory divides telecommunications into four major parts: fixed voice telecoms, fixed datacoms, mobile voice telecoms, and mobile datacoms. At the end of 2001 their sizes in traffic were fixed datacoms, fixed voice, mobile voice, and mobile datacoms. At the same time the sizes of revenues changes the order as fixed datacoms does not bring proportionately nearly as much revenue as fixed and mobile voice, hence the order by revenues starting from the largest is fixed voice, mobile voice, fixed datacoms, mobile datacoms.

17.1.1 Historical trends

Fixed voice was, for all practical purposes, the only form of telecommunications 15 years ago. By year 1990 fixed voice dominated telecommunications with call types of local calls, long distance calls and international calls each dwarfing the small revenue streams from fledgling fixed datacoms and mobile voice.

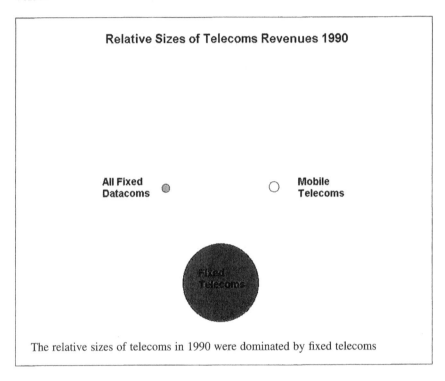

Relative Sizes of Telecoms Revenues 1990

All Fixed
Datacoms

Mobile
Telecoms

Fixed
Telecoms

The relative sizes of telecoms in 1990 were dominated by fixed telecoms

Fixed voice traffic and revenues continued to grow from 1990 to 1995. Fixed data communications emerged as another type of telecommunications, and was used to link data networks. During the early 1990s first Internet e-mail and then the web browser started to build datacoms traffic and the fixed datacoms traffic started to grow dramatically. Yet its revenues were not as big as the traffic. The cellular telephone networks were still mostly on analogue systems and by 1995 the revenues from mobile telecoms were not particularly large either. But both fixed datacoms and mobile voice had become significant elements in global telecoms.

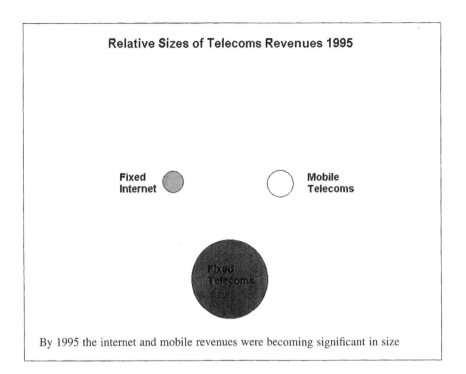

Relative Sizes of Telecoms Revenues 1995

Fixed Internet

Mobile Telecoms

Fixed Telecoms

By 1995 the internet and mobile revenues were becoming significant in size

From 1995 to 2000 fixed voice traffic and revenues continued to grow at a gradual pace. The fixed datacoms evolved into what we call Internet traffic today. By 2000 fixed data traffic had exceeded fixed voice traffic worldwide, but the revenues from fixed data still did not match fixed voice. Mobile voice telecoms went digital during the mid- to late 1990s and while its traffic is less than datacoms traffic, mobile voice tariffs are so much greater that the mobile voice revenues already accounts for more revenues worldwide than fixed datacoms. Year 2000 saw the newest element in the picture emerge as mobile data traffic, growing fast both in traffic and revenues.

17.1.2 Fixed voice calls

The oldest and still in 2001 the biggest by revenues, is fixed voice telecommunications. Fixed voice includes all local voice calls, long distance and international calls. Fixed voice call traffic and fixed voice call revenues have grown steadily over the past 10 years. Compared with the other major types of telecommunications traffic, the growth rates in fixed voice have been the most gradual.

17.1.3 Fixed datacoms/fixed Internet

Fixed datacommunications first emerged to allow early data networks to connect to each other. With the rapid development in the personal computer industry, as modems became faster and especially after the advent of web browsers, fixed datacommunications traffic grew extremely fast. Total world-wide fixed datacommunications traffic already exceeds total worldwide fixed voice traffic. But typical of fixed datacoms traffic, especially narrowband Internet traffic and its considerable price competition and various 'free access' services, the revenues of fixed datacoms have always been consider-ably less than the relative volume of traffic. Thus fixed datacoms revenues have grown a lot slower than their overall traffic.

17.1.4 Mobile voice telecoms

Mobile voice telecoms became a significant traffic form with the advent of cellular telecommunications. Early predictions for the usage of cellular phones had maximum theoretical user amounts at levels of 10% of the population, and even second generation (digital) cellular phone usage esti-mates held theoretical maximums at about 20% penetrations. In 1998 Finland became the first advanced telecoms market where total mobile penetration exceeded fixed penetration, and now phone penetration esti-mates expect cell phone penetrations to level off at about 120–130%. This means that most people between age 8 and 80 will have mobile phones, and a good deal of the working age population will have two, sometimes three phones.

As the amount of users grew, so too grew traffic. And with mobile commu-nications, a new telecommunications concept emerged, that of Reachability. Reachability and mobile telecoms traffic patterns are discussed in more detail in the chapter on Patterns (Chapter 11). As mobile phone calls are more expensive than their equivalent fixed voice calls, the duration of mobile calls tend to be shorter than fixed calls. Often mobile calls are very short 'Where are you' and 'I will be there shortly' type of updates. Mobile call traffic has grown very rapidly over the past 10 years, but their revenue growth has been much more dramatic.

17.1.5 Mobile datacoms/mobile Internet

The newest type of service of the four analysed here, is mobile Internet (mobile datacoms) traffic. For all practical purposes mobile Internet traffic did not exist before 1998. By 2001 the mobile Internet traffic was generating already revenues well in excess of a billion dollars worldwide. More significantly, the mobile Internet traffic growth is exhibiting the same usage growth patterns as the fixed Internet traffic did in the 1990s, and better yet, the mobile Internet traffic revenue growth is illustrating similar growth as mobile voice did in the 1990s. Understandably most analysts have identified the mobile Internet revenues as the fastest growing segment of the overall telecoms market.

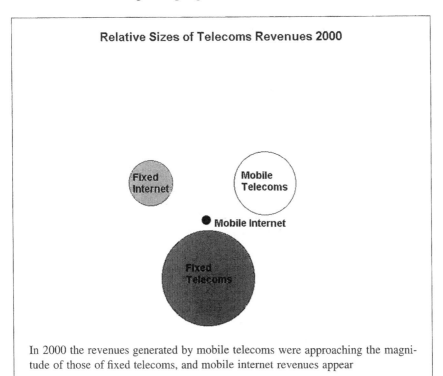

Relative Sizes of Telecoms Revenues 2000

Fixed Internet

Mobile Telecoms

Mobile Internet

Fixed Telecoms

In 2000 the revenues generated by mobile telecoms were approaching the magnitude of those of fixed telecoms, and mobile internet revenues appear

17.1.6 Projecting into the future

When we project the growth paths into the future, it is reasonable to assume that worldwide fixed telecoms traffic and revenues will continue to grow at a gradual pace even through 2005. As fixed datacoms traffic has grown at dramatic pace over the past 10 years, it is reasonable to assume that it will still grow fast through 2005. The Internet industry has been trying to generate stronger revenues from the business, it is reasonable to assume that revenues in fixed data will grow as well. Mobile voice traffic and revenues have grown fast especially over the last 5 years and are likely to grow for the next 5 years. The mobile Internet only appeared in 1998 and had already grown to over 1 billion dollars worldwide by end of 2000. It is reasonable to assume that it will continue to grow the fastest and as it generates a lot of revenues per traffic when compared with the fixed Internet, soon the point will come when the mobile Internet will generate as much revenues as the fixed Internet.

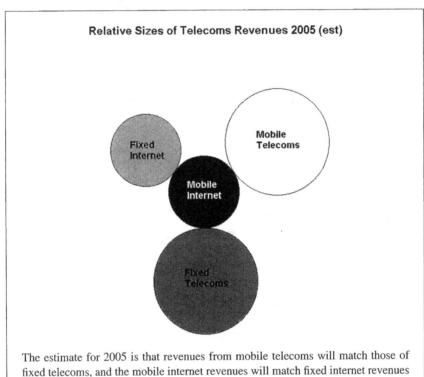

The estimate for 2005 is that revenues from mobile telecoms will match those of fixed telecoms, and the mobile internet revenues will match fixed internet revenues

My prediction of the rough proportions of the traffic revenues worldwide is that around 2005 the worldwide fixed voice revenues will be matched by worldwide mobile voice revenues. This may happen even earlier. Also roughly at the same time the fixed Internet revenues will be matched by mobile Internet revenues. This may happen a little later, depending on the take-up of GPRS (General Packet Radio System), EDGE (Enhanced Data Rates for Global Evolution) and 3G. The larger pair by revenues will be fixed and mobile voice, and the smaller pair will be fixed and mobile Internet. But within a few years two pairs of roughly similar size elements will emerge. Many analysts have made predictions that are similar to or in harmony with these, it is more about the exact timing of when will mobile voice revenues catch fixed voice revenues, etc., that is under discussion; nobody doubts that it will happen before the decade is done. There are those who then say, but how much money is there in the overall economy, we cannot move all of the spending to telecoms. That type of macroeconomic shift in the economy has also been examined and most analysts agree that overall spending on telecoms will increase somewhat over the next 5 years as JP Morgan for example shows.

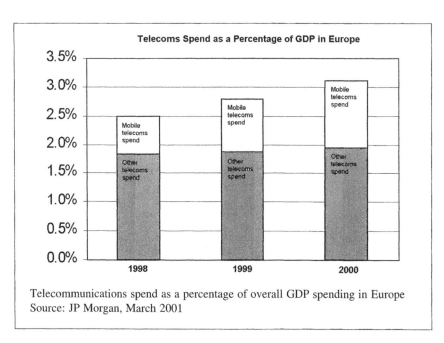

Telecommunications spend as a percentage of overall GDP spending in Europe
Source: JP Morgan, March 2001

The situation roughly in year 2005, with the two pairs of equally sized telecoms revenue groups, is the premise upon which my theory of Money Migration is based. I will not be discussing how each of these groups will grow, but rather what migration of revenues will happen between the four elements as we approach 2005.

17.2 Migration and fixed voice

Fixed voice traffic will continue to grow into the near future. There is a considerable amount of voice tariff erosion, but most analysts feel that the voice tariff erosion is less than overall voice traffic growth, thus overall fixed voice revenues are likely to grow into the next few years.

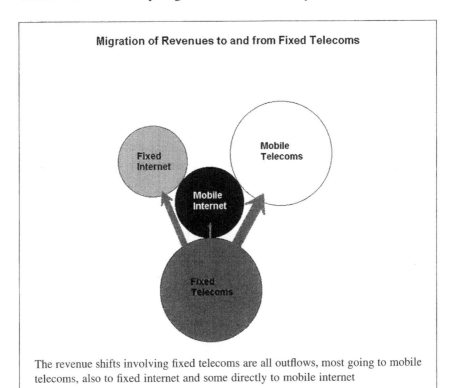

Migration of Revenues to and from Fixed Telecoms

The revenue shifts involving fixed telecoms are all outflows, most going to mobile telecoms, also to fixed internet and some directly to mobile internet

17.2.1 Migration from fixed voice

There is considerable voice traffic cannibalisation already happening. This cannibalisation is somewhat lost in the analysis and as an issue, due to the continued growth of overall fixed voice traffic.

There are several types of shifts. The first and most known migration is fixed mobile substitution. It does not mean that people will stop using the fixed phone altogether, although such trends are clearly visible in Scandinavia among younger adults. It means that of all voice traffic, ever more voice calls are placed on mobile phones rather than on fixed phones, and also, that ever more frequently people call others on their mobile phones rather than calling their fixed phones. As a very significant migration, ever more of voice calls start off, end on, or are completely on mobile phones. This results in voice revenues shifting from fixed voice to mobile voice. As this has been happening for a while several studies have also documented it, such as Deutsche Bank which reported on Japanese calling patterns.

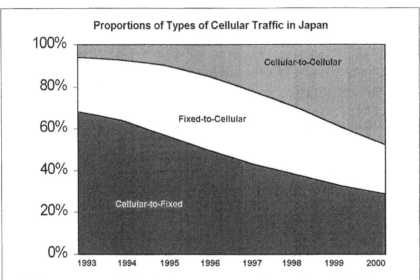

Traffic patterns are shifting, in Japan ever more calls are from mobile phone to another mobile phone,
Source: Deutsche Bank Wireless Internet Report, May 2001

A second migration is happening simultaneously. Many of the existing fixed voice connections are still being used, but no longer for significant voice traffic, but rather for fixed Internet traffic. Again it does not have to be exclusively fixed Internet, but a clear shift is happening where the existing home phone connection is ever more likely to be connected to a computer modem. If families upgrade to ISDN (Integrated Services Digital Network) or ADSL (Asymmetric Digital Subscriber Line), the reason for this in almost all cases is the need to get faster Internet connections even if they still maintain also a voice connection. This migration of what was exclusively fixed voice traffic and revenue, is towards increasingly more fixed Internet/datacoms traffic.

A third migration is very likely to occur, in smaller scale, over the next 5 years. It is the migration of traffic and revenues directly from fixed voice to mobile Internet. This would be typical for the late adopter type of user or family, or for example a family which gets a sudden need to connect to the Internet. Or it could be a new family with small children, who would be Internet-aged in the next few years. In the near future the handheld mobile Internet devices will become compelling in purchase price advantage over buying a PC to access the Internet. At that time, if a family was still not connected to the Internet and finds a need to do so, it is likely it would not get a fixed Internet connection and an Internet-enabled PC, but rather a mobile Internet device and jump directly to the mobile Internet fraternity. This money migration is likely to be much smaller than the two others affecting fixed voice, but it should be considered as well.

17.2.2 Migration to fixed voice

The prospect of migration from the other parts back into fixed voice needs to be considered. It is a very unlikely scenario that anybody would abandon a mobile phone and insist on going back to using fixed voice only. In very extreme economic conditions this might happen, for example if a working person decides to go back to university and all at once finds disposable income to be drastically reduced and wants to limit expenses. Even in that case it is quite unlikely that the person would totally abandon the mobile phone, but rather that the person would put more of the voice traffic on to fixed voice rather than mobile voice. For the big picture one could argue that there will be no migration back from mobile voice to fixed voice.

Even less likely is migration from fixed Internet to fixed voice. It is very unlikely that a user who is accustomed to accessing the Internet would stop using it and revert to only using fixed voice calls. One should keep in mind that this analysis is of *users* and their revenues, not of *households*. So if a family had a computer for their son, and the son moves away from home and takes the computer with him, and now that family has no longer a need for fixed Internet traffic in the home, for the economy, the boy will still use his Internet, only from his new home. No money migrated back to fixed voice, even if the family fixed phone line is now used only for fixed voice calls.

Thus there is no money migrating back to fixed voice. A lot of money is migrating out of fixed voice, mostly to mobile voice, also significant amounts to fixed Internet, and probably also soon small amounts directly to mobile Internet.

17.3 Migration and mobile voice

We saw that there is migration from fixed voice toward mobile voice. We also saw that there is no return migration from mobile voice to fixed voice. A significant other migration affecting mobile voice is that towards the mobile Internet. After a user has become accustomed to the mobile phone, it is one device which is no longer given up as we saw in the early chapters of this book. But new features can be added to the mobile phone. SMS (Short Message Service) text messaging is an early example of new functionality that can be added to the mobile phone. And SMS text messages are of course the most primitive form of mobile Internet traffic. Since then the Japanese experiences with DoCoMo's I-Mode, and KDDI's and J-Phone's experiences with the Japanese variation of WAP (Wireless Application Protocol) have shown that more advanced services can be rapidly adopted by the mobile phone user.

It should be noted that mobile voice traffic will not be migrating to fixed Internet traffic. If a mobile phone user starts to use the fixed Internet, it will be *in addition to* the current mobile phone spending, not as a substitute for

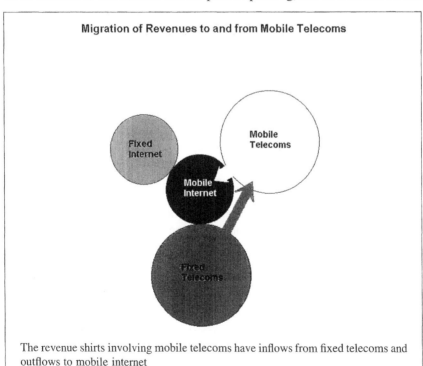

Migration of Revenues to and from Mobile Telecoms

The revenue shirts involving mobile telecoms have inflows from fixed telecoms and outflows to mobile internet

it. But mobile Internet spending can be a substitute for mobile voice, such as in many cases of SMS usage where it is the preferable means of contact when a phone call might be too intrusive. One should also keep in mind that often mobile data contact results in voice calls, such as people responding to an SMS text message by calling the person who sent the message, or in more advanced mobile Internet services the ability to 'click to talk' and establish instant voice contact.

Mobile voice traffic and revenues get an infusion from fixed voice, and cannibalisation by mobile Internet. But no migration goes the other way, and no migration exists between mobile voice and fixed Internet.

17.4 Migration and fixed Internet

We've seen already that fixed Internet/fixed datacoms gains some revenue migration from fixed voice, and does not get any meaningful revenue migration from mobile voice. There is a small additional revenue stream coming as migration from the mobile Internet. This would be some of those people who have never had a fixed Internet connection, and whose first Internet experience is via the mobile Internet device. Some of these people will like the Internet but not like the small handheld device and will want to gain faster speeds, large displays and other such convenience and user experience. Such people will migrate some, but probably not all of their mobile Internet traffic from the mobile Internet to the fixed Internet.

A significant and very likely greater money migration will be that from the fixed Internet to the mobile Internet. As the mobile Internet experience gets more user friendly, many who are hooked to the fixed Internet and would like to take it with them everywhere, will like the mobility and always-on aspects of the mobile Internet and start to migrate some of their traffic from the fixed Internet to the mobile Internet.

The migration from fixed Internet to mobile Internet will also be true for personal computer equipment and fixed line subscriptions which become obsolescent. For example if the home computer is an old 486 class computer with old Internet software running over Windows 3.1, when the time comes to upgrade, if a new PC costs in the magnitude of 1000–2000 dollars, and a mobile Internet-enabled phone costs between 300 and 400 dollars, a strong economic argument emerges to upgrade from the PC to the mobile Internet phone (or PDA – Personal Digital Assistant). Furthermore, early research suggests that when given similar services, users prefer mobile services over fixed ones. Therefore, it is reasonable to assume that while money migrates in

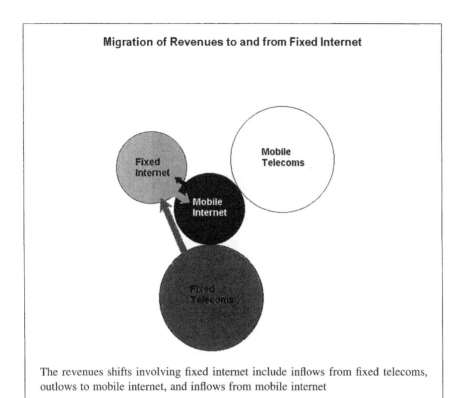

Migration of Revenues to and from Fixed Internet

The revenues shifts involving fixed internet include inflows from fixed telecoms, outlows to mobile internet, and inflows from mobile internet

both directions, the direction from fixed Internet to mobile Internet is likely to be bigger. Definitely there will not be fixed Internet users abandoning the Internet altogether and only use mobile voice or fixed voice. So no money migrates in the directions of mobile voice or fixed voice.

17.5 Migration and mobile Internet

The last of the four groups of services is the youngest of the four, mobile Internet services. All of the revenue streams were discussed in the previous three analyses and shown here.

As we've already seen, there is a money stream from fixed voice directly to mobile Internet. This is likely to be relatively small. There is no stream going from mobile Internet to fixed voice. There will be a large stream of money migrating from mobile voice to mobile Internet and again no money migrating the other way. And we've seen that there will be a relatively large flow of

Migration of Revenues to and from Mobile Internet

The revenues involving mobile internet include inflow from fixed telecoms, mobile telecoms and fixed internet. There is an outflow to fixed internet

money migrating from fixed Internet to mobile Internet, but a counterbalancing, probably smaller stream of money migrating from mobile Internet to fixed Internet.

17.5.1 Conclusion of money migration

Each of the four groups will continue to grow both in traffic and revenues. Therefore it may be difficult to notice that there are significant shifts in traffic, users and revenues. These bring about opportunity for profits for the players involved in telecoms.

Over the next 5 years or so, the fixed voice is a group which only loses revenues in Money Migration. Money migrates to each of the other three. Mobile voice gains from fixed voice, but loses to mobile Internet. Fixed Internet gains from fixed voice and mobile Internet and gives revenues back to mobile Internet. The mobile Internet gains from all three but also gives revenues back to fixed Internet.

Theory of Money Migration in Telecoms in the Near Future

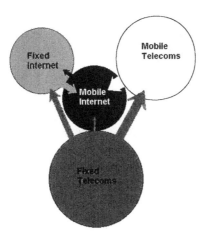

The theory of Money Migration shows the shifts in revenues among the four major types of telecoms traffic

17.6 Relevance of Money Migration

The relevance of the theory comes down to opportunities. If one considers any single service type by itself, it will be expected to grow. Services which offer more fixed voice services to fixed voice users will be familiar business serving a rather well-known market and customer, amongst familiar and entrenched competition. If you examine the money migration stream, you find change in behaviour, and opportunity for new services and innovation. In such instances there are likely to be less competitors, and those that exist are likely to be still in the learning stages of that marketplace. This is where it is easy to make a major success and to satisfy considerable customer needs. If you consider the direction against the flow, there is likely to be very little opportunity and small markets if any. Here it will be very difficult to make significant traffic as well as significant profits.

Companies, services, applications, recruitment, etc. should be keenly aware of the money migration in the industry. The magnitude of the money migration is hidden beneath the large growth trends in all four forms of communications. Individuals should seek career opportunities with companies, partners, departments, bosses, etc. who work with the money migration flows, not against them or irrelevant to them. That brings customers, satisfaction, profits, bonuses, raises, promotions, etc. The money migration areas will also allow for service innovation meaning that many exciting new services and ideas – and very significantly those which are likely to survive in the marketplace and make money – will emerge from money migration streams.

17.7 Changes

This chapter has shown what are the great currents of money in the evolving flows of revenues in telecommunications. This chapter illustrates opportunities and thus highlights where easy growth, success, profit and promotion can be had. Some will not grasp the significance of this chapter, but others might understand what Winston Churchill meant, when he said: ''Men occasionally stumble over the truth, but most of them pick themselves up and hurry off as if nothing has happened.''

18

'The wise man will make more opportunities than he will find.'

Roger Bacon

4G:

What next?

3G or UMTS (Universal Mobile Telecommunications System) is a terrestrial mobile telecoms network. It is based on a standard called IMT-2000 for the next generation of telecoms networks after the first digital networks which were called second generation. The analogue cellular networks were called first generation. If UMTS is called 3G, then it is very reasonable to assume there will be a fourth generation, and also that some discussion about this future be included in this book.

18.1 What is 4G

Following the media hype relating to the unanticipated costly auction licences for 3G in the UK and Germany, the topic of 4G emerged and we find many definitions of what 4G is. This book takes the replacing network principle, that 2G actually replaced 1G networks, and any enhancements to first generation networks were not 2G. Similarly enhancements to 2G are so-called 2G + or 2.5G services, but as long as the second generation network can be utilised, it is an evolution of that generation, and only when a wholly new network is needed, is it third generation. Thus standards such as GPRS (General Packet Radio System) and EDGE (Enhanced Data Rates for Global Evolution) are enhancements to 2G, and thus 2.5G or 2G + systems, not 3G systems, even if they are capable of running 3G-type services. There is another view which sees 3G as a performance improvement over 2G, and any wireless technology which has better performance than 3G would be 4G. That may be a reasonable position on service performance, but not on network technology, as it does not assume a new *generation* of cellular networks. The G in 3G is, after all, a generation, not a speed indicator. The generations from first to third were illustrated in Kaaranen *et al.*'s book *UMTS Networks*.

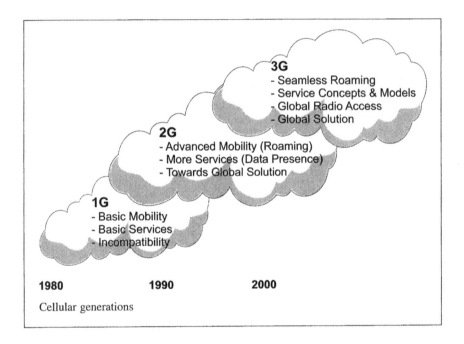

Cellular generations

Today, with the standards for WCDMA (Wideband Code Division Multiple Access) and CDMA2000 and other 3G systems, these are expected also to be gain enhancements and supplementary systems and solutions. Some may talk about these as next generation systems or 4G systems. Equally, some who have not paid for a licence may try to use the interest in what is next, i.e. 4G, and use that interest to create market awareness of simply complementary solutions to 3G, while arguing that these are 4G solutions. Typical in this type of argument is that those players claim great operational efficiencies and technical superiority over the 'older' 3G solutions.

That type of argument may hold water if considering two separate techno-logical solutions to one problem – like considering long distance travel and comparing jet aeroplanes to bullet trains. Either can be argued to be better in some conditions.

But it would be unfair to say that the next generation of aeroplanes, after the jumbo jet generation, would be bullet trains. That could arguably be the next evolution of travel, but as long as trains cannot cross oceans, then a train cannot replace a significant part of air travel, and thus not be a total replace-ment solution. This brings back the discussion to what is 3.5G and possibly 4G.

There are several emerging technologies which allow for wireless commu-nication in efficient and high data speed ways, which exist today or will exist very soon. They may provide much of the utility of 3G and in some conditions can be clearly superior to 3G. Arguably a train can be many times more efficient than a jet aeroplane, for example in the local traffic of one part of downtown to another – such as the subway trains in most major cities. A jet aeroplane, although able to transfer hundreds of passengers, would be too inefficient and noisy, and require remarkable take-off and landing space – to compete against a subway train in moving people from one part of the city to another.

Similarly several new wireless solutions can be used to provide partial data and voice transmissions, and often at better speeds or lower costs than 3G.

18.2 3.5G – or enhancements possible with 3G networks

Enhancements to the 3G network from various supplementary, supporting, and partially competing technologies, which can coexist with 3G, will be called 3.5G services for the purposes of this book. It covers Bluetooth, W-LAN (Wireless Local Area Network) and digital TV. Some have called solutions built upon these technical platforms as 4G.

18.2.1 Bluetooth

The first and most obvious is Bluetooth. This standard allows for very short-range wireless communication between devices. Bluetooth ranges are measured in metres or feet, not in kilometres or miles. Bluetooth is an excel-lent solution to connect laptop computers to printers and large screen projec-tion screens, etc. Bluetooth transmission is also expected to be so cheap to manufacture that Bluetooth is likely to be installed in a very wide variety of devices and gadgets and even clothing. For example Bluetooth could be the way a portable CD player communicates with a wireless headset or earphones. Bluetooth could be the way that a laptop computer could wirelessly connect to a mobile phone in a nearby computer bag, etc.

Bluetooth is envisioned to allow many very local connections to happen. Typically these would be in areas where the connected device can be seen, often even touched. So for example if the mobile phone would need to talk to a vending machine to transfer money so that you could buy a coke, that could happen via the mobile network – e.g. via SMS (Short Message Service) text messaging. Or it could happen locally via Bluetooth. This could be 'free of charge' short-range connecting.

Similarly you might walk into a hotel and a sensor in the hotel doorway would 'talk' to your mobile phone via Bluetooth and find out who you are and guide you to your pre-ordered room, unlock doors, etc. Bluetooth is an excellent means for very short-range connections.

18.2.2 Wireless LAN

Wireless LAN (also known as 802.11 or WiFi) is the evolution of LAN technology from what was once on coaxial cabling, then on 'twisted pair' telephone cabling, and which has had an evolution towards wireless solutions being developed for at least 10 years. Wireless LAN allows computers to connect to each other, and to shared peripherals such as printers and servers, without cabling. Wireless LAN was intended initially as an office solution and its costs and performance have been mostly compared with delivering the same LAN performance over cabled networks.

Wireless LAN found a niche for itself in telecommunications as more and more laptop computer users were connecting to their corporate LANs via cellular phone-based modems at airports, hotels and conference centres. Soon these institutions noticed that by providing W-LAN connections to the travellers, they were serving them and allowed the laptop computer users much faster access to their intranets, e-mails and the Internet.

18.2.3 Hot spots

Now several content providers are building business solutions to deliver content via W-LAN, either as a direct competitor to the 2.5G and 3G services offered by the mobile operators, or as a complement to them. The concept is called hot spots. You might have a hot spot for example at a petrol service station and while you fill your car with petrol, you can also download data to your car, such as music to your music system or the latest newspapers, or games for the kids and a book for your wife to read on the journey. Hot spots can also be used by stores owned by the content provider, the most obvious example being Disney stores and the like. So Disney might make its electronic content available upon subscription at a website or 3G service provider, and then offer it at a discounted rate directly from their store, via W-LAN.

There are many concerns with W-LAN on issues such as billing and security. For the actual use on a handheld device, the device needs to have the W-LAN feature built into the device, adding cost and weight, or connected via adapter or accessory, adding cost, bulkiness, inconvenience and risk of technical compatibility conflict. Still W-LAN will definitely be a part of the wireless services future.

18.2.4 Virtual hot spots

The prudent 3G operator will plan for the emergence of W-LAN and join in the creation of hot spots. The 3G operator could use its experience and market presence to make itself one of the major players in the W-LAN area and thus be part of that competition, perhaps even dominating the space. The 3G operator could also create 'virtual hot spots' and provide a 'simple man's wireless LAN' without W-LAN cards and technology. The 3G network operator will be having a lot of excess capacity early in the 3G network build-up phase, and some of this excess capacity could be used to create virtual hot spots, where *designated data services* could be offered for nearly free and at high speeds. So rather than wait for W-LAN operators to come and set up their systems at the airports, hotels and conference centres, the 3G network operator could offer the service at those locations for example by simply dialling a special access code. This would be a defensive marketing move.

18.2.5 Digital TV data broadcasts

Another future area of development will utilise digital TV transmissions. The digital TV transmission ability has dramatic power for simulcasting data. It is not a convenient solution to handle individual communication between handheld devices. But on items of mass data delivery, such as the whole issue of today's daily newspaper with colour images, etc., the data volume is easily in the hundreds of megabytes. Or to put it in another way, it would eat up a data

CD disk. To transmit that amount on the 3G network would consume remarkable amounts of resources for a long time. But digital TV has the ability to set channels to transmit data, and if so, this type of data quantity might be transmitted in seconds or minutes, to all who have subscribed to the service.

Digital TV will not replace 3G networks, for the simple reason that digital TV cannot handle voice calls and other communication between individuals efficiently. But digital TV will probably be introduced either as a compliment or a competitor – or both – for some of the high data volume traffic that would be consumed on mobile handsets. In that case the multi-purpose mobile phone would also need a digital TV receiver, or at least parts of it, which adds to the complexity and cost of the handset.

18.3 4G and four visions

What will then take over after 3G networks have become overcome by traffic congestion and more capacity is needed. Here are brief views into four possible visions.

18.3.1 Terrestrial radio network

The safest bet for the future of mobile communications is to assume it will follow a natural progression from the current. It was similar but younger radio engineers who progressively specified 2G after 1G, and 3G after 2G. There is a lot of creativity, innovation and invention happening in radio transmission technology, miniaturisation, data compression, etc. Currently both the WCDMA and CDMA2000 evolution paths include significant performance improvements in their specifications for the near future. It is quite likely that after enough development has happened, that in some years a new technical standard could be developed allowing for much higher bandwidth data transmissions. It is likely that by that time there would be no distinction between voice and data, voice would have evolved into just one data application on the radio networks. A logical name for the next version of UMTS could be BMTS as in *Broadband* Mobile Terrestrial System. And much like broadband services in fixed telecoms, xDSL (Digital Service Lines) are much faster than their earlier cousins, such as ISDN (Integrated Services Digital Network), so too should the new mobile system be faster than the fastest conceivable evolution of 3G, by at least a factor of four, hopefully more such as a factor of 10.

18.3.2 Satellites

While early satellite communication systems fell prey to the simultaneous emergence of 2G terrestrial systems, which were able to deliver communications at a fraction of the cost of satellite systems, one should not count out the satellite option. Just like in aviation the Concorde and Tupolev 144 were the first passenger jets to travel at twice the speed of sound, they emerged at exactly the same time as the jumbo jets which cut the cost of transporting a single person to a fraction of what it was when the supersonics were designed. While the Tupolev is out of service and Concorde struggled with its profitability for years before the accident which grounded the fleet, man still is not happy to fly long-haul flights that last 12–15 h, such as crossing the Pacific. Constantly newer airliner designs emerge proposing supersonic aircraft, and even faster ones. Eventually they will come of course.

I am not suggesting that a satellite-based telecommunications system would automatically be faster or cheaper than a terrestrial system, but rather that the satellite industry is only taking its first steps in the game, and they may emerge as strong candidates to deliver '4G'. Many proposals exist to increase the throughput of satellite-based systems.

18.3.3 Peer-to-peer mobile networks

One futuristic vision of what will be the form of the next mobile network, is that a similar evolution would happen in mobile networking as happened in data networking with the Internet. The idea is that each mobile terminal could start to act as both a terminal and a base station, or to use datacoms terminology, that each mobile terminal would act in essence as a mobile router. Then there would be no – or at least much less – need for expensive base stations and their antennae. Each device would act as a chain to continue the message. Standardising work is being done to explore this possibility as a design for the next evolution of mobile networking. Certainly very many technical and commercial issues would need to be solved before this type of network could emerge as commercially viable, and even further until they could start to replace the 3G networks now coming on line. But peer-to-peer networking was the very first steps that enabled local area networking for personal computers – which then in turn was a key element in enabling the Internet revolution in fixed telecommunications. Peer-to-peer might do that in mobile communications as well.

18.3.4 Digital TV data broadcast evolution

The digital TV broadcast technology is excellent for delivering simultaneous data transmissions to masses. Currently much of what we use our mobile phone for, is person-to-person traffic, and for that digital TV is not a good option. But if we take another analogy from IT, the mirroring of Internet sites, then one could see a potential for a future where the majority of the traffic on 4G networks would be data. As such the network would be optimised for data, not voice. For data, the majority of the data could be data-heavy content such as games, video clips, music, etc., and the person-to-person communications could be a small subset of the total traffic. If so, then a solution built upon the efficient delivery of mass quantities of data, and digital TV would be suitable for that, could become the basis for 4G. In that case the issues of individual person-to-person communication would still need to be solved, and that could even perhaps be done with a combination of 3.5G technologies.

Other proposals for 4G exist. At this point in late 2001 we do not know for sure what standards will be implemented and what will eventually win out as 4G. It should be noted that the first standardising work for 3G was completed in the early 1990s which resulted in IMT-2000. From that standard it took the telecoms equipment manufacturers another 10 years to be able to deliver radio transmission equipment and the very first mobile phone handsets. So even if 4G was delivered twice as fast and its standardisation could be completed in a few years, there is at least a window of 10 years of 3G before any 4G network can even be deployed.

18.4 Tomorrow never dies

And as we saw with the migration from 1G and 2G in the 1990s, and will see from 2G to 3G in the early 2000s, it takes again years for the newer standard to replace the existing one. So the current 3G network operators can rather safely assume that 4G will not cannibalise their 3G service revenues for at least a decade.

But if the other parts of this book have been relatively clear visions of a very near future, this 4G chapter is definitely much more hazy about the more distant future. So I need to go back in history and seek guidance from someone who ushered in a new industrial age. Henry Ford said this about the future: "There is only one thing in business that is certain and that is change. I don't know what tomorrow is going to be like but I do know this: its bound to be different from yesterday and today."

19

'The next time you are in a meeting, look around and identify the yesbutters, the notnowners and the whynotters. Whynotters move companies.'

General Electric advertisement in 1984

Postscript:

Final thoughts

It is the latest evolution of the Information Age, the Computer Age, the Networked Computer Age, the Mobile Age, the Converged Age, the Connected Age. We have the traditional fixed Internet and the new mobile Internet. Very soon there will be more connected devices on the mobile Internet than on the fixed Internet.

When the majority of the web surfers will arrive at content not from PCs but from mobile terminals, then also the mobile content will migrate to mobile environments to serve those surfers. That transition is inevitable. The sheer numbers dictate that direction, although it is made even faster by the fact that it is easy to bill for content on the mobile Internet, and near impossible to do so on the fixed Internet. It is likely, that with the opportunities to make money much easier on the mobile Internet than on the fixed Internet, the content providers may prefer the mobile Internet and migrate their services to the mobile Internet much faster than we now expect. But sooner or later, whether in 2004 or 2006, it will definitely happen.

For the content provider the issue is crystal clear. As one media provides money to you and the other does not, it is clear where the best content will first emerge and where serviced innovations will be introduced. That is the time when the mobile Internet will be found to have won over the fixed Internet. Perhaps then we will have a natural sounding term for the mobile Internet, something like the 'm-net' or the 'wireless web' to distinguish the discussion from the (old-fashioned and fixed) Internet. When the content migrates, that brings about the opportunity for content creators and owners to be paid for their intellectual property.

The mobile phone has already become part of everybody's personal gear, taken along everywhere and kept close at all times. This connectedness has brought about the immediacy, reachability and ability to communicate always and with everyone. We are building new communities with our mobile phones. 3G services will greatly enhance that ability to communicate, to access information and entertainment, and to be connected. Services that assist in that community aspect will be seen as valuable.

In that world where multiple connectedness becomes the norm and most people are connected somehow, it becomes ever more important to be able 'to network' or to build personal networks and manage those relationships in new and efficient ways. Earlier when the industrial age was turning into the information age, information and control of it became increasingly valuable. When information was collected and processed, but not delivered for further handling, *controlling* information was power. Modern management theory is already ridiculing the information-hoarding dinosaurs in miscellaneous business bureaucracies. In e-business the sharing of information is vital and those who are most efficient at sharing information are the ones with the new power. The information age is now changing into the mobile Internet converged age. The connected age. In the connected age, even more than in the information age *sharing* information is power.

The 3G mobile operators will be at the heart of the new converged services world. Their actions, more than those of any others, will determine how healthy this industry will be, and how soon it will reach the trillion dollar mark. The network operators will be uniquely able to influence the market interest and adoption of preferred access devices and services.

Network operators cannot do it alone. They are facing more competitive pressure than ever before, are taking a giant leap into a new technology, release a vast array of new services, and at the same time they have to learn to partner with numerous content and application partners. Some will be fast learners. Others will stumble and learn from their mistakes. But in the scramble for 3G success, all the players should keep in mind that market share is not the answer, *profitability* is.

This book has looked at the new mobile services from their revenue and profit points of view. The book has also introduced several practical theories to help in service creation, revenue projections, and money migration in the industry.

Hopefully some of the services outlined in this book have helped identify what kind of 3G services could be created, how they can become profitable propositions, and give some insight who are natural partners in those situations. Hopefully the brief descriptions of what might be the money-sharing mechanism will serve to start up the dialogue between the parties who have to partner.

This book is an early attempt to look at some services that are likely to emerge in the 3G environment over the next few years. Some of the services in this book are probably going to be universal and commonplace within a few years, others may be seen as strange curiosities which failed in the marketplace. Of course hundreds of exciting new services will emerge which were never mentioned in this book. What those will be only time can tell.

This book has discussed the money aspect of 3G services and where billions of dollars of money will be made over the first decade of the new millennium. That future holds incredible opportunities for those who are astute and take advantage of it. Hopefully many who will work in new mobile Internet services can identify with Goethe who said: ''It is not doing the thing we like to do, but liking the thing we have to do, that makes life blessed.''

I hope that this book has inspired you.

'Laughter is the cheapest luxury a businessman has. It stirs up the blood, expands the chest, electrifies the nerves, clears away the cobwebs from the brain, and gives the whole system, a general rehabilitation.'
E.C. McKenzie

m-Dictionary:

Humorous interlude

Soon the letter m will dominate the business theory landscape not unlike the e letter does today with e-mail, e-business, e-commerce, etc. So your m-Tomi now proudly presents, the following submissions into the vocabulary of business English.

m-Action – the behaviour of calling somebody's voice mail or leaving a SMS (Short Message Service) text message, and doing nothing else, to create the illusion of having done something.

m-Ahonen – a mildly amusing original piece of spam e-mail comedy, with a strong sense of déjà vu and links to previous pieces. *"What was in the e-mail, dear? Oh nothing, just an m-Ahonen."*

m-Bailing – the act of disconnecting from a live mobile game when one is losing, and blaming it on poor network coverage. *"I can't stand playing against Del, whenever he's losing, he m-Bails."*

m-Barter – the process of having the poor phone user (who pays his own bill) switch with the rich phone user (whose employer pays phone bills) so that the rich phone user calls and places the cost of the call on to his employer. *"Let me call you back, as my employer pays for my phone bills."*

m-BC – (MBC) the latest all-news channel for yet another smaller targeted audience launched by NBC and CNBC after they decided they have not lost enough money in their current channels.

m-Boss – the kind of boss who is never physically present but always barks orders at you via the cellular phone.

m-Break – taking an unexpected moment out of a hectic day to chat with a long-lost friend who just happens to call you from a holiday to some exotic place – like Rio, Goa or Hawaii.

m-Broke – not having credit on the pre-paid phone account (see m-Barter). *"I can't call you – I'm m-Broke."*

m-Climax – the most intense type of anticlimax when an intense and emotional moment is ruined by the ringing of the mobile phone, such as the opening night of a Broadway play or passionate sex with a new lover.

m-Depression – feeling of loneliness when nobody sends you an SMS. *"I called you because I was having a bad case of m-Depression."*

m-Friend – the person you never see anymore but still talk, chat or exchange SMS with. *"First we were colleagues, then we were lovers, then we became friends; but now we're only m-Friends."*

m-Four – (M4) a motorway in England (oops, wrong dictionary, this should be in the catalogue of British Roads and Highways).

m-Grammr – the nu abbrvtd language of SMS txt msgs.

m-Heckling – to make fun of someone via SMS text messaging as the person speaks. *"Would the audience please turn their mobile phones to silent when m-Heckling the speakers."*

m-M – the next boss of James Bond, a virtual reincarnation of the original M.

m-M&A – mergers and acquisitions in the mobile operator/cellular carrier world.

m-M&M – smaller m&ms which are more mobile. The green ones taste the best.

m-mm – the mobile millimetre, a tiny measure of distance best described by that change of a cellular network connection from perfect bars to loss of connection.

m-Mobile – (Mobile Mobile) the likely name of some new pop band in line with Mr Mister, Wet Wet Wet, Tony Toni Tone, etc.

m-Moment – taking moment to accept a call in middle of discussion. *"I can't believe that right in the middle of the job interview he took an m-Moment."*

m-N-m – (eminem) the new hip, nastier and more mobile version of the rap star Eminem.

m-NOPQ – the new mobile alphabet used in mobile phones (see m-Grammr). *"Today class we study the m-NOPQ for your mobile phone exam next week."*

m-Overkill – how the m prefix will soon dominate marketing and be over-exposed just like the e prefix today. Its likely successor as the latest omni-present prefix to be n for New or Nu. *"The m-Overkill actually started from an innocent comedy piece, would you believe?"*

m-Pathy – listening to someone's problems via mobile phone and not having to do anything else; almost like caring.

m-Rejection – not wanting more SMSs when your phone memory is already full of all the great messages you want to keep forever.

m-Stale – the way this m-joke parallels the previous e-joke by the same (lame) m/e-Ahonen.

m-Sync – a newer younger boy band following in steps of nSync.

m-Tee – the empty feeling one feels after getting brushed off in romantic pursuit via SMS rather than in person.

m-Thumb Syndrome – (MTS) a repetitive action injury to the fingers when having to type the same number keys multiple times to get letters, especially prone to hit young SMSers.

m-Time – a new concept of time of frequent occurrence and indefinite duration, situated between soon and now, symptomised by SMS text messages and calls from mobile phones, giving estimates of arrival. *"I am almost there."*

m-Tomi – your mobile comic-wannabe.

m-TV – (mTV) MusicTV's new mobile Internet version.

m-Void – the gulf of separation which kills all relationships which try to sustain themselves over long periods of time only via mobile phone connection. *"What happened, you two were so close? Oh, he just flew away and our relationship sank into the m-Void."*

m-Wakeup – using the alarm of the mobile phone to wake up rather than an unreliable hotel wakeup call.

m-Yelling – the way people increase their speaking volume when speaking on a mobile phone.

The above for entertainment purposes only. Does not reflect the opinions of any of the employers, present nor past, nor current customers and colleagues of the author, nor any of the companies or public personalities mentioned. Original humour by Tomi T Ahonen/HatRat 2001/2002.

'Why is abbreviate such a long word?'
George Carlin

Abbreviations

0-1-2-3	(Service Creation Aid): 0 manuals, 1 button internet, 2 seconds maximum delay, 3 keystrokes maximum to reach service
1G	First Generation mobile networks
2G	Second Generation mobile networks
3G	Third Generation mobile networks
4G	Fourth Generation mobile networks
5 M's	Movement Moment Me Money Machines
ADSL	Asynchronous Digital Subscriber Line
AMPS	Advanced Mobile Phone System
ARPU	Average Revenue Per User
ASP	Application Service Provision
ATM	Automatic Teller Machine
B2B	Business to Business
B2C	Business to Consumer
B2E	Business to Employee
BBS	Bulletin Board System
CAPEX	CAPital EXpenses
CD	Compact Disk
CDMA	Code Division Multiple Access

CDR	Call Detail Record / Charge Detail Record
CEO	Chief Executive Officer
CRM	Customer Relationship Management
DJ	Disc Jockey
DOS	Disc Operating System
DVD	Digital Video Disk
EDGE	Enhanced Data rates for GSM Evolution
FM	Frequency Modulation
FTP	File Transfer Protocol
GDP	Gross Domestic Product
GPRS	General Packet Radio System
GSM	Global System for Mobile communications
HiFi	High Fidelity
HQ	Head Quarters
HSCSD	High Speed Circuit Switched Data
HTML	HyperText Markup Language
ID	Identity
IMT-2000	International Mobile Telephony (2000)
IP	Internet Protocol
IRC	Internet Relay Chat
ISDN	Integrated Services Digital Network
ISP	Internet Service Provider
IT	Information Technology
kbps	kilobits per second
LAN	Local Area Network
mAd	Mobile Advertising
MAGIC	Mobile Anytime Globally Integrated Customized
Mbps	Megabits per second
m-DJ	Mobile (network) Disc Jockey
MIDI	Musical Instrument Digital Interface
MP3	Motion Picture Experts Group Audio Layer 3
MPEG	Motion Picture Experts Group
MTV	Music TV
MVNO	Mobile Virtual Network Operator
NMT	Nordic Mobile Telephone
OPEX	OPerating EXpenses
PAIR	Personal Available Immediate Real time
PC	Personal Computer
PDA	Personal Digital Assistant

QoS	Quality of Service
R&D	Research and Development
ROM	Read Only Memory
RAN	Radio Access Network
SIM	Subscriber Identity Module
SIP	Session Initiation Protocol
SME	Small and Medium Enterprise
SMS	Short Message Service
SOHO	Small Office Home Office
TDMA	Time Division Multiple Access
UMTS	Universal Mobile Telecommunications System (also Universal Mobile Telecommunication Services)
VCR	Video Cassette Recorder
VHE	Virtual Home Environment
VPN	Virtual Private Network
WAN	Wide Area Network
WAP	Wireless Application Protocol
W-ASP	Wireless Application Service Provider
WCDMA	Wideband Code Division Multiple Access
WiFi	Wireless Fidelity
W-LAN	Wireless Local Area Network
WWW	World-Wide Web

'No furniture so charming as books.'
Sydney Smith

Bibliography

Accenture: Future of wireless, Cambridge: 2001

Ahonen T, Barrett J: Services for UMTS, Chichester: Wiley, 2002, 373 pp

Ahonen T, Kasper T, Melkko S: 3G Marketing, Chichester: Wiley, 2002, 340 pp

Comer DE: Internetworking with TCP/IP, volume I: Principles, protocols and architecture. 4th edition. Englewood Cliffs, NJ: Prentice Hall, 2000, 755 pp.

Frengle N. i-Mode, a primer: M&T Books, New York, 2002, 485 pp

Halonen T, Melero J, Romero J: GSM, GPRS & EDGE performance. Chichester: Wiley, 2002, 614 pp

Hannula I, Linturi R: 100 phenomena, virtual Helsinki and the cybermole (translation from Finnish by William More) e-book at http://www.linturi.fi/100_phenomena/ Helsinki: Yritysmikrot 1998 212 pp

Holma H, Toskala A: WCDMA for UMTS, revised edition. Chichester: Wiley, 2001, 313 pp

IDATE: Web Music: Issues at stake and forecasts. Montpellier 2001

JP Morgan: Mobile Matters. London 2001

Kaaranen H, Ahtiainen A, Laitinen L, Nahgian S, Niemi V: UMTS networks. Chichester: Wiley, 2001, 302 pp

Kalakota R: M-Business. Columbus: McGraw Hill, 2001, 272 pp

Kopomaa, T: City in your pocket, the birth of the information society. Helsinki: Gaudeamus, 2000, 143 pp

Laiho J, Wacker A, Novosad T: Radio network planning and optimisation for UMTS. Chichester: Wiley, 2001, 512 pp

Lamont D: The Age of m-Commerce. Oxford: Capstone, 2001, 288 pp

May P: Mobile commerce. Cambridge: Cambridge University Press, 2001, 302 pp

May P: The business of E commerce. Cambridge: Cambridge University Press, 2000, 288 pp

McLelland S: Ultimate telecom futures. Guilford: Horizon House, 2002, 232 pp

Nokia: Mobile Messaging. Espoo 2001

Nokia: Mobile Entertainment. Espoo 2001

Ojanperä T, Prasad R: Wideband CDMA for third generation mobile communications. Boston, MA: Artech House, 1998, 439 pp

Ovum: Consumer Portals. London 2001

Sadeh N: M-Commerce. Chichester: Wiley, 2002, 272 pp

Stallings W: Data & computer communications. Englewood Cliffs, NJ: Prentice Hall, 2000, 810 pp

Tennent J, Friend G: Economist guide to business modelling, London: Economist Books 2001, 272 pp

UMTS Forum, Report 16: 3G portal study, available at the UMTS Forum website http://www.umts-forum.org/reports_r.html November 2001

UMTS Forum, Report 13: Structuring the service opportunity, available at the UMTS Forum website http://www.umts-forum.org/reports_r.html April 2001

UMTS Forum, Report 12: Naming, addressing and identification issues for UMTS, available at the UMTS Forum website http://www.umts-forum.org/reports_r.html February 2001

UMTS Forum, Report 11: Enabling UMTS third generation services and applications, available at the UMTS Forum website http://www.umts-forum.org/reports_r.html October 2000

UMTS Forum, Report 10: Shaping the mobile multimedia future, available at the UMTS Forum website http://www.umts-forum.org/reports_r.html October 2000

UMTS Forum, Report 9: UMTS third generation market structuring the service revenue opportunities, available at the UMTS Forum website http://www.umts-forum.org/reports_r.html October 2000

Wilkinson N: Next generation network services. Chichester: Wiley, 2002, 216 pp

Webb G: The m-bomb. Oxford: Capstone, 2001, 256 pp

'In a connected age sharing information is power.'

Tomi T Ahonen

Useful Websites

160 characters (SMS and Mobile Messaging Association)
http://www.160characters.com/

3GPP (Third Generation Partnership Project)
http://www.3gpp.org/

Accenture
http://www.accenture.com

ALACEL (Latin American Wireless Industry Association)
http://www.alacel.com/home.cfm?lang=en

ARC Group
http://www.arcgroup.com

ARIB (Association of Radio Industries and Business) (Japan)
http://www.arib.or.jp/index_English.html

Boeing
http://www.boeing.com

BWCS
http://www.bwcs.com/

Chalmers University of Technology, Göteborg Sweden
http://www.chalmers.se/Home-E.html

Compwise
http:www.compwise.com

CTIA (Cellular Telecommunications & Internet Association)
http://www.wow-com.com/

CWTS (China Wireless Telecommunication Standards Group)
http://www.cwts.org/cwts/index_eng.html

Deutsche Bank
http://www.db.com/

ETSI (European Telecommunications Standards Institute)
http://www.etsi.org/

FCC (Federal Communications Commission)
http://www.fcc.gov/

GBA (Global Billing Association)
http://www.globalbilling.org

GSM Association
http://www.gsmworld.com

GSA (Global mobile Suppliers Association)
http://www.gsacom.com/

GSM Association
http://www.gsmworld.com

IDATE
www.idate.fr

IEEE (Institute of Electrical and Electronics Engineers)
http://www.ieee.org/

IETF (Internet Engineering Task Force)
http://www.ietf.org

ITU (International Telecommunications Union)
http://www.itu.ch/

John Wiley & Sons
http://www.wiley.com/

JP Morgan
http://www.jpmorgan.com/

MDA (Mobile Data Association)
http://www.mda-mobiledata.org/

MEF (Mobile Entertainment Forum)
http://www.mobileentertainmentforum.org/

MGIF (Mobile Gaming Interoperability Forum)
http://www.mgif.org/

MMA (Mobile Marketing Association) - Note: formed out of merger of WAA and WMA
http://www.waaglobal.org/

Nokia
http://www.nokia.com

OFTEL (Office of Telecommunications) (UK)
http://www.oftel.gov.uk/

Ovum
http://www.ovum.com/

PayCircle
http://www.paycircle.org/

PCIA (Personal Communications Industry Association)
http://www.pcia.com/

PPA (Periodicals Publishing Association) UK
http://www.ppa.co.uk/

SyncML
http://www.syncml.org/

T1 (Committee T1)
http://www.t1.org/

The 3G Portal
http://www.the3Gportal.com

Tilastokeskus / Statistics Finland
http://www.tilastokeskus.fi/index_en.html

TomiAhonen.Com
http://www.tomiahonen.com

TTA (Texas Telephone Association)
http://www.tta.org/

TTC (Telecommunication Technology Committee) (Japan)
http://www.ttc.or.jp/e/

UMTS Forum
http://www.umts-forum.org/

UWCC (Universal Wireless Communications Consortium)
http://www.uwcc.org/

Verista
http://www.verista.com

WAA (Wireless Advertising Association) - Merged with WMA to form MMA, see MMA above

WAP Forum
http://www.wapforum.org

WCA (Wireless Communications Association International)
http://www.wcai.com/

WISPA (Wireless Internet Service Providers Association)
http://www.wispa.org/

WLANA (Wireless LAN Association)
http://www.wliaonline.com/

WLIA (Wireless Location Industry Association)
http://www.wliaonline.com/

WMA (Wireless Marketing Association) - Merged with WAA to form MMA, see MMA above

Yankee Group
http://www.yankeegroup.com/

*There are risks and costs to a program of action
but they are far less than the long-range risks
and costs of comfortable inaction.'*

John F Kennedy

Index to Services

> 'What great thing would you attempt if
> you knew you could not fail?'
> Robert Schuller

Index *(Services In Italics)*

Printed and bound by CPI Group (UK) Ltd, Croydon, CR0 4YY

23/04/2025

14660946-0001